Nabil Benoudjit

Variable Selection and Neural Networks

Nabil Benoudjit

Variable Selection and Neural Networks

Application in Infrared Spectroscopy and Chemometrics

VDM Verlag Dr. Müller

Imprint

Bibliographic information by the German National Library: The German National Library lists this publication at the German National Bibliography; detailed bibliographic information is available on the Internet at http://dnb.d-nb.de.

Cover image: www.purestockx.com

Publisher:
VDM Verlag Dr. Müller Aktiengesellschaft & Co. KG , Dudweiler Landstr. 125 a, 66123 Saarbrücken, Germany,
Phone +49 681 9100-698, Fax +49 681 9100-988,
Email: info@vdm-verlag.de

Zugl.: Louvain-la-Neuve, UCL, Diss., 2003

Produced in USA and UK by:
Lightning Source Inc., La Vergne, Tennessee, USA
Lightning Source UK Ltd., Milton Keynes, UK
BookSurge LLC, 5341 Dorchester Road, Suite 16, North Charleston, SC 29418, USA

ISBN: 978-3-8364-9504-2

To my father

Acknowledgments

First of all, I would like to express my most sincere gratitude to my advisor, Prof. Michel Verleysen, for his solid technical guidance, continuous support and encouragement in my work, his permanent availability and responsiveness to my requests and for his friendship and hospitality.

I would like to thank the Secrétariat à la Coopération Internationale of the UCL for its financial support for about 48 months.

Of course, I am also very grateful to all the members of the laboratory of spectrophotomery of the research unit AGRO/BNUT of UCL, and more particularly to Prof. Marc Meurens and Ir. Etienne Cools, who have shared their knowledge and experiences with me.

Working with the Artificial Neural Networks Group members was very helpful and pleasant. I would like to thank all the members of the Artificial Neural Networks Group: John Aldo Lee, Nicolas Donckers, Amaury Lendasse, Cédric Archambeau, Damien François, Geoffroy Simon and Frédéric Vrins for their collaboration, help and very useful comments on all my work during four years.

Contents

List of Figures

11

List of Tables

Chapter 1

Introduction

1.1 Context and motivations

The infrared spectra of agricultural and food products contain information which presents an analytical interest. However, the extraction of this information is not immediate and requires almost always a rather complex mathematical treatment. Indeed, the spectra are the result of an interaction of light with matter which one cannot completely describe from a theoretical point of view.

There are many definitions of chemometrics. One of the most frequent states the following [33]:

Definition: Chemometrics is a new chemical branch which uses the theory and the methods developed in statistics, mathematics and computer sciences to extract useful and substantial information from chemical measurements. Another term used in the literature for chemometrics is multivariate analysis.

This book focuses particularly on the application of chemometrics in the field of analytical chemistry. Chemometrics (or multivariate analysis) consists in finding a relationship between two groups of variables, often called dependent and independent variables. In infrared spectroscopy for instance, chemometrics consists in the prediction of a quantitative variable (the obtention of which is delicate, requiring a chemical analysis and a qualified operator), such as the concentration of a component present in the studied product, from spectral data measured on various wavelengths or wavenumbers (several hundreds, even several thousands).

We distinguish two operations:

1. modelling in laboratory where all measurements of variables (dependent and independent variables) must be carried out and where parameters of the model (linear or non-linear) must also be estimated,

2. using only the measured new independent variables to predict the dependent one, once the parameters of the model are estimated (see figure 1.1).

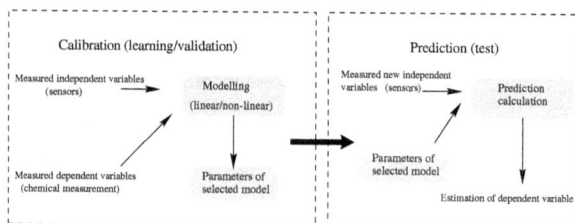

Figure 1.1: Calibration and prediction diagram block of the spectrophotometric data.

From a chemometric point of view, the spectrum obtained in infrared spectroscopy is a complex function that depends on both the physical and chemical properties of the sample [15], and have remarkable characteristics which require specific methods for their treatment.

The spectrophotometric data may often comprise more independent variables (spectral data) than observations (spectra or samples). This case is rather less encountered in other applications of statistics. Collinearity of the independent variables is typical for spectrophotometric data, i.e. certain independent variables can be practically represented as a linear combination of other independent ones; this is the source of many problems in direct application of many statistical methods, such as the Multiple Linear Regression (MLR) [5, 13, 27, 31, 54]. Studies have shown that if collinearity is present among the variables, the prediction results can get poor (see for example [27, 79]). This limitation has promoted other alternative linear methods to offset the problems generated by the strong redundancy between variables. Several alternatives that are able to adapt to this collinearity were developed, that are Stepwise Multiple Linear Regression (SMLR) [13, 23, 38, 55, 62], Principal Component Regression (PCR) [31, 34, 40, 54, 55] and Partial Least Square Regression (PLSR) [13, 34, 40, 44, 54, 55, 80, 88], etc.

In analytical chemistry, a lot of linear calibration methods (quoted above) are applied to solve quantitative determination problems with the argument that the relation between the chemical composition and the measured signal is linear [20]. However, there are many situations where non-linearity is present. For instance, Miller [59] discusses important sources of non-linearity in near-infrared spectroscopy, namely

- deviations from the Beer-Lambert law, which are typical of highly absorbing samples;

- non-linear detector responses;

- drifts in the light source;

- interactions between analytes;

- non-linearity between diffuse reflectance/transmittance data and chemical data.

When the non-linearity is significant, one can use truly non-linear calibration techniques, e.g. Artificial Neural Networks (ANN).

To summarize, spectral data obtained from spectrophotometers have the following characteristics:

1. great number of spectral data (several hundreds, even several thousands),

2. more spectral data (variables) than spectra (observations),

3. high collinearity between spectral data,

4. non-linear relationship between the spectral data (independent variables) and the analyte concentration (dependent variable).

1.2 Objectives of the book and our way to solutions

The objective of this work is to predict the concentration (dependent variable) of analyte present in a studied product from independent variables, which are spectral data measured on various wavelengths or wavenumbers. Since the spectrophotometric data have specific characteristics quoted in the previous section, it is necessary to select variables among the candidates, in order to build a still suitable model with only few variables. The objective of the variable selection is three-fold: improving the prediction performances, providing faster and more cost-effective prediction, and providing a better understanding of the underlying process that generated the data [39]. In this work we propose a methodology in the field of chemometrics to handle the spectrophotometric data which are often represented in high-dimension. To handle these data, we first propose a new incremental method (step-by-step) for the selection of spectral data using linear and non-linear regression based on the combination of three principles: linear or non-linear regression, incremental procedure for the variable selection, and use of a validation set. This procedure allows on one hand to benefit from the advantages of non-linear methods to predict chemical data (there is often a non-linear relationship between dependent and independent variables), and on the other hand to avoid the overfitting phenomenon, one of the most crucial problems encountered with non-linear models. Secondly, we propose to improve the previous method by a judicious choice of the first selected variable, which has a very important influence on the final performances of the prediction. The idea is to use a measure of the mutual information between the independent and dependent variables to select the first one; then the incremental method (step-by-step) is used to select the next variables. The variable selected by mutual information can have a good interpretation from the spectrochemical point of view, and does not depend on the data distribution in the training and validation sets. On the other hand the traditional chemometric linear methods such as PCR or PLSR produce new variables which do not have an obvious interpretation from the spectrochemical point of view.

The non-linear calibration techniques used in this work are Artificial Neural Networks (ANN). Why using Artificial Neural Networks? Because these are powerful tools that may be used in a wide variety of applications, when non-linear modelling of data is needed. The true power and advantages of ANN lie in their ability to represent both linear and non-linear relationships, and in their ability to learn these relationships directly from the data being modeled. Traditional linear models are simply inadequate when it comes to modelling data that are non-linearly related.

In this work we have chosen to use Radial Basis Functions Networks (RBFN), because they can be used for a wide range of applications and their training is faster compared to the popular Multilayer Perceptrons (MLP) [46]. This fast speed comes from the fact that RBFN have just two layers of parameters which can be determined sequentially. RBFN allow the modelling of non-linear data using a linear approach, which is therefore fast, with the additional benefit of avoiding the problem of local minima usually encountered when using MLP [45]. Besides, while the available literature on Radial Basis Function Networks learning widely describes how basis function centers and weights must be set, little effort has been devoted to the learning of basis function widths. For this reason, the book addresses this topic: it shows the importance of a proper choice of basis function widths, and how inadequate values can dramatically influence the approximation performances of the RBF Networks.

1.3   Organization of the book

The book is divided into two parts. The first part aims at introducing the Radial Basis Functions Networks and the impact of the width of the Radial Basis Functions on the obtained solution. The second part is devoted to the spectrophotometric variable selection.

Part I: Artificial Neural Networks (ANN)

Chapter 2 presents a broad overview of Radial Basis Function Networks (RBFN). It also covers alternative ways of training RBFN, and the impact of the width of the radial basis functions on the obtained solution. In this chapter we also propose a heuristic to optimize the widths in order to improve the generalization process.

Chapter 3 gives some general procedures that have been widely used in this work for model selection.

Chapter 4 gives a theoretical example and three analytical ones that are used to evaluate the heuristic proposed in chapter 2. This chapter shows the importance of a proper choice of the widths of the Gaussian kernels (radial basis functions), and how inadequate values can strongly influence the approximation performances of the RBF Networks.

Part II: variable selection

Chapter 5 addresses the problem of the spectrophotometric variable selection. Two variable selection methods are presented. Firstly, we present a new incremental method (step-by-step) for the selection of spectral data (variables) using linear and non-linear regression (the latter being implemented by Radial Basis Function Networks). The heuristic proposed in chapter 2 is used to perform the learning of RBFN. Secondly, we propose a sequential method to distribute data homogeneously between the training and validation sets. Thirdly, we suggest a way to improve the previous incremental procedure by a better choice of the first variable, which has an important effect on the final performances of the modelling. Four real-life datasets are presented in order to show the efficiency and advantages of both proposed procedures compared to the traditional chemometric linear methods often used, such as MLR, PCR and PLSR.

The last chapter gives a general synthesis of the research developed in this book and discusses some ideas for future work.

The main contributions of the book are:

1. with respect to Artificial Neural Networks ([6, 11]):

   - a one-dimensional searching procedure as a compromise between an exhaustive search on all basis function widths, and a non-optimal a priori choice.

2. with respect to variable selection ([7, 8, 9, 10]):

   - a new variable selection method for spectral data using linear and non-linear regression (the latter being implemented by RBF Networks);

   - a data distribution method between training and validation sets;

   - an improved method for the spectrophotometric variable selection using non-linear methods (mutual information and RBF Networks).

Part I

Artificial neural networks

Chapter 2

Radial Basis Function Network (RBFN)

2.1 Introduction

Artificial Neural Networks (ANN) are widely used in applications involving classification or function approximation. It has been proven that several classes of ANN such as Multilayer Perceptron (MLP) and Radial-Basis Function Networks (RBFN) are universal function approximators [14, 65, 66]. Therefore, they are widely used for function approximation [66].

Radial-Basis Function Networks (RBFN) can be used for a wide range of applications, primarily because they can approximate any function and their training is faster compared to Multilayer Perceptrons (MLP) [46]. This fast learning speed comes from the fact that RBFN have just two layers (figure 2.2) of parameters (centers + widths and weights) and that each layer can be determined sequentially. MLP are trained by supervised techniques: the weights are computed by minimizing a non-linear cost function. On the contrary the training of RBF networks can be split into an unsupervised part and a linear supervised part. Unsupervised updating techniques are straightforward and relatively fast. Moreover, the supervised part of the learning consists in solving a linear problem, which is therefore also fast, with

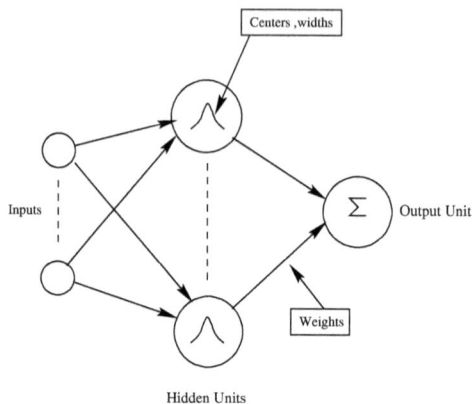

Figure 2.1: Architecture of a Radial Basis Function Neural Network.

the additional benefit of avoiding the problem of local minima usually encountered when using multilayer perceptrons [45]. The training procedure for RBFN can be decomposed naturally into three distinct stages: (i) locating the centers of the radial functions, (ii) determining their widths (usually standard deviations of Gaussian kernels), and (iii) calculating the network weights between the radial functions layer and the output layer.

This chapter presents a broad overview of Radial Basis Function Networks (RBFN), and can help to understand their properties by using concepts from approximation theory. It also covers several aspects with immediate practical implications: alternative ways of training RBFN, and the impact of the width of the radial basis functions on the obtained solution. Several algorithms and heuristics are available in the literature regarding the computation of the centers of the radial functions [2, 37, 71] and the weights [14, 63]. However, very few works only are dedicated to the optimization of the widths of the Gaussian kernels. In this chapter, we develop a heuristic to optimize the widths in order to improve the generalization process.

2.2 Radial basis function network

A RBF network is a two-layered ANN. Consider an unknown function $f(\mathbf{x}) : R^d \to R$. RBF networks approximate $f(\mathbf{x})$ by a weighted sum of d-dimensional radial activation functions (plus linear and independent terms). The radial basis functions are centered on well-positioned data points, called centroids. The centroids can be regarded as the nodes of the hidden layer. Usually, the positions of the centroids and the widths of the radial basis functions are obtained by an unsupervised learning rule. The weights of the output layer are calculated by a supervised process using pseudo-inverse matrices or singular value decomposition (SVD) [14]. However, it should be noted that other authors use the gradient descent algorithm to optimize the parameters of RBF network [81]. The training strategies of RBF networks will be detailed in next section.

Suppose we want to approximate a function $y = f(\mathbf{x})$ with a set of M radial basis functions $\phi_j(\mathbf{x})$, centered on the centroids $\mathbf{c}_j$ and defined by:

$$\phi_j : R^d \to R : \phi_j(\mathbf{x}) = \phi_j(\| \, \mathbf{x} - \mathbf{c}_j \, \|), \qquad (2.1)$$

where $\| \, . \, \|$ denotes the Euclidean distance, $\mathbf{c}_j \in R^d$ and $1 \leq j \leq M$.

The approximation of the function $y = f(\mathbf{x})$ may be expressed as a linear combination of the radial basis functions [67]:

$$\hat{y} = \hat{f}(\mathbf{x}) = \sum_{j=1}^{M} \lambda_j \phi_j(\| \, \mathbf{x} - \mathbf{c}_j \, \|) + \sum_{i=1}^{d} a_i x_i + b, \qquad (2.2)$$

where λ_j are weight factors, a_i and b_i are the weights for the linear and independent terms respectively.

A typical choice for the radial basis functions is a set of multi-dimensional

Gaussian kernels:

$$\phi_j(\| \; \mathbf{x} - \mathbf{c}_j \; \|) = exp\left(-\frac{1}{2}(\mathbf{x} - \mathbf{c}_j)^T R^{-1}(\mathbf{x} - \mathbf{c}_j)\right), \qquad (2.3)$$

where $\mathbf{c}_j$ is the centre and R is the metric. The term $(\mathbf{x} - \mathbf{c}_j)^T R^{-1}(\mathbf{x} - \mathbf{c}_j)$ is the distance between the input $\mathbf{x}$ and the centre $\mathbf{c}_j$. Often the metric is Euclidean. In this case two options can be considered:

1. $R = \sigma^2 I$, where I is the identity matrix and σ is a common scalar width for all Gaussian kernels. In the literature, several authors used this option [21, 43, 64, 66, 71]

2. $R = \sigma_j^2 I$, where σ_j is a scalar width of the jth Gaussian kernel. For example, in [51, 60, 70, 82], each scalar width σ_j is estimated independently.

In this chapter, we use a scalar width covering options (1) and (2), while general covariance matrices R can be used too [61]. It should be noted that the use of covariance matrices to adjust the widths of each Gaussian kernel is very sensitive to outliers [69], and many authors (see for example [46]) suggest to use option (2) for simplicity reasons.

2.3 RBFN learning strategies

The selection of appropriate number of basis function M is related to the important problem of model order selection, and is treated in detail in chapter 3. Once the number and the general shape of the radial basis functions $\phi_j(\mathbf{x})$ are chosen, the RBF network has to be trained properly. Given a training dataset T of size N_T,

$$T = \{(\mathbf{x}_p, y_p) \in R^d \times R, 1 \le p \le N_T : y_p = f(\mathbf{x}_p)\}, \qquad (2.4)$$

the training algorithm consists in finding the parameters $\mathbf{c}_j$, σ_j, λ_j, a_i and b such that $\hat{y}$ fits the unknown function y as close as possible. This is realised by minimizing a cost function (usually the normalized mean square error between $\hat{y}$ and y on the learning points). Often, the training algorithm is decoupled into a three-stage procedure:

1. determining the centers $\mathbf{c}_j$ of the Gaussian kernels,

2. computing the widths σ_j of the Gaussian kernels,

3. computing the weights λ_j and independent terms a_i and b.

During the first two stages only the inputs $\mathbf{x}$ of the training dataset T are used. The parameters are thus adapted according to an unsupervised updating rule. In the third step the weights λ_j, independent terms a_i and b are calculated with respect to the corresponding desired outputs; meanwhile $\mathbf{c}_j$ and σ_j remain fixed. Furthermore, values found through the three-stage procedure may be adjusted in a subsequent phase: all parameters are optimized according to gradient descent on the cost function (normalized mean square error). The global optimization by descent gradient looses the advantages of split learning, but usually requires only a few iterations of global optimization, which reinforces the computational advantages of RBFN over MLP-like networks. Note that a direct gradient descent on all parameters after random initialization is usually not used, because it suffers from the same drawbacks as MLP: greater risk of local minima, flat regions in the error function leading to inefficient gradient descent, and the need of complex optimization algorithms, etc... [83]. The main attraction of RBF network is its three-stage procedure training, which is lost while using the global optimization.

2.3.1 Location of centroids

Locations of centroids are usually chosen according to the density of the input set $\mathbf{x}_i$; such a choice leads to more centroids, and so naturally to a better approximation of targets y_i, in regions of the input space covered by more input vectors, which seems to be a good heuristic in many applications.

Moody and Darken [60] use the K-means clustering algorithm, in which the number K of centers must be decided in advance. Suppose we need to partition N data points $\mathbf{x}^n$ into K clusters and find the corresponding cluster centers. The K-means algorithm seeks to partition the data points into K subsets S_j by minimizing the sum of squares clustering function [41]:

$$J = \sum_{j=1}^{K} \sum_{n \in S_j} \| \mathbf{x}^n - \mathbf{c}_j \|^2, \tag{2.5}$$

where $\mathbf{c}_j$ is the mean of the data points in the subset S_j and is given by

$$\mathbf{c}_j = \frac{1}{N_j} \sum_{n \in S_j} \mathbf{x}^n. \tag{2.6}$$

One method of finding these clusters is by using the batch version. First the data points are randomly assigned to K subsets. The centers for each of the subsets are then computed. The data points are then reassigned to the cluster whose centre is nearest. This procedure is repeated until there is no further change in the grouping of the data points. Other authors use a stochastic online process (Competitive Learning) method, which leads to similar results, with the advantage of being adaptive (continuous learning, even with an evolving input database). The principle is as follows:

1. to initialise the M centroids by a random choice in the training data set;

2. to use recursively all data points $\mathbf{x}_p$, and move the closest centroid $\mathbf{c}_j$ to data points $\mathbf{x}_p$ (Best Matching Unit, or BMU) according to

$$\mathbf{c}_j(t+1) = \mathbf{c}_j(t) + \alpha(t)(\|\mathbf{x}_i - \mathbf{c}_j\|), \tag{2.7}$$

where $\alpha(t)$ is a time decreasing adaptation factor, $0 < \alpha(t) < 1$.

After convergence of this Competitive Learning procedure, the density of the centroids will approximate the density of the input data. Other unsupervised techniques such as Kohonen's self organizing feature map can also be used for determining the centroids [50].

2.3.2   Width factors

The second stage of the training process involves the computation of the Gaussian function widths, while fixing the degree of overlapping between the Gaussian kernels. This allows finding a compromise between locality and smoothness of the function $\hat{y}$. We consider here both options (1) and (2) quoted in section 2.2. Option (1) consists of taking identical widths $\sigma_j = \sigma$ for all Gaussian kernels [21, 43, 64, 66, 71]. In [43] for example, the widths are fixed as follows:

$$\sigma = \frac{d_{max}}{\sqrt{2M}} \qquad (2.8)$$

where M is the number of centroids and d_{max} is the maximum distance between any pair of them. This choice would be close to the optimal solution if the data were uniformly distributed in the input space, leading to a uniform distribution of centroids. Unfortunately most real-life problems show non-uniform data distributions. The method is thus inadequate in practice and an identical width for all Gaussian kernels should be avoided.

If the distances between the centroids are not equal, it is better to assign a specific width to each Gaussian kernel. For example, it would be reasonable to assign a larger width to centroids that are widely separated from each other and a smaller width to closer ones [46]. Option (2) therefore consists of estimating the width of each Gaussian kernel independently [51, 60, 70, 82]. This can be done, for example, by splitting the learning points $\mathbf{x}_p$ into clusters according to the Voronoi region associated to each centroid[1], and then computing the standard deviation of the distance between the learning points in a cluster and the corresponding centroid; reference [82], for example, it is suggested to use an iterative procedure to estimate this standard deviation. Moody and Darken [60], on the other hand, proposed to compute the width factors σ_j (the radius of kernel j) by the p-nearest neighbours heuristic:

---

[1]A voronoi region is the part of the space nearest to a specific centroid than to any other one.

$$\sigma_j = \frac{1}{p} \left(\sum_{i=1}^{p} \parallel \mathbf{c}_j - \mathbf{c}_i \parallel^2 \right)^{\frac{1}{2}}, \qquad (2.9)$$

where the $\mathbf{c}_j$ are the p-nearest neighbours to centroid $\mathbf{c}_i$. A suggested value for p is 2 [60]. Saha and Keeler [70] proposed to compute the width factors σ_j by the nearest neighbour heuristic where σ_j (the radius of kernel j) is set to the Euclidean distance between $\mathbf{c}_j$ (the vector determining the centre for the jth RBF) and its nearest neighbour $\mathbf{c}_i$, multiplied by an overlap constant r :

$$\sigma_j = r \cdot min \left(\parallel \mathbf{c}_j - \mathbf{c}_i \parallel \right). \qquad (2.10)$$

This second class of methods offers the advantage of taking the distribution variations of the data into account. In practice, they are able to perform much better than fixed-width methods, as they offer a greater adaptability to the data. Even though, as we will show next in chapter 4 with examples, the width values given by the above rules remain sub-optimal.

2.3.3   Width Scaling Factor optimization (our approach)

We suggest in this book a procedure for the computation of the Gaussian function widths based on an exhaustive search belonging the second class of algorithms quoted in subsection 2.3.2, the purpose being to show the importance of the optimization of Gaussian widths. Therefore we select the widths in such a way to guarantee a natural overlap between Gaussian kernels, preserving the local properties of RBFN, and at the same time to maximize the generalization ability of the network.

First we compute the standard deviations σ_j^c of the learning data in each cluster in a classical way [6, 11].

Definition:
Sigma-cluster is the empirical standard deviation of the learning data

contained in a cluster or Voronoi region associated to a centroid.

Subsequently, we determine a width scaling factor WSF, common to all Gaussian kernels. The widths of the kernels are then defined as:

$$\forall j, \ \sigma_j = WSF \cdot \sigma_j^c \qquad (2.11)$$

Equation 2.11 offers a compromise between the usual methods without optimization of σ_j and a M-dimensional optimization of all σ_j together, which would be computationally much too complex and expensive.

By inserting the width factor WSF, the approximation function $\hat{y}$ is smoothed such that the generalization process is possibly improved, and an optimal overlapping of the Gaussian kernels is allowed. Unfortunately, the optimal width factor WSF depends on the function to be approximated, the dimension of the input set, as well as on the data distribution. The choice of the optimal WSF value is thus obtained by extensive simulations (cross-validation): the optimal WSF_{opt} value will be chosen as the one minimizing the error criterion (mean square error on a validation set), among a set Q of possible WSF values.

When several minima appear, it is recommended to choose the one corresponding to the smallest width scaling factor (see figure 2.2). Indeed, large WSF have to be avoided for complexity, reproducibility and/or numerical instability. When widths are allowed to be large, the localized receptive field concept becomes inapplicable. This will be illustrated in the chapter 4, in which we prove the effectiveness of our approach on several artificial problems.

2.3.4 Optimal weights

Once the basis function parameters are determined, the transformation between the input data and the corresponding outputs of the hidden units is fixed. The network can thus be viewed as an equivalent single-layer network with linear output units. The output is calculated by a linear combination (i.e. a weighted sum) of the radial basis function plus the

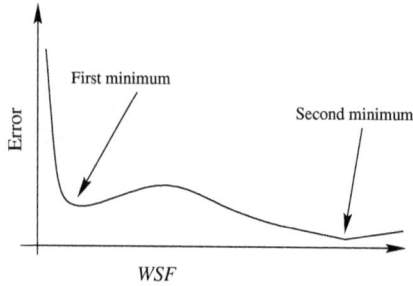

Figure 2.2: Curve of the error according to *WSF*.

independent terms.

$$W = \varphi^+ \mathbf{y} = (\varphi^T \varphi)^{-1} \varphi^T \mathbf{y}, \qquad (2.12)$$

where $W = [\lambda_1 \ \lambda_2 \ \ldots \ \lambda_k \ a_1 \ a_2 \ \ldots \ a_d \ b]^T$ is the column vector of λ_j weight factors, independent term a_i and b, and $\mathbf{y}$ is the column vectors of y_p training data outputs.

Moreover

$$\varphi = \begin{pmatrix} \phi_{11} & \phi_{12} & \cdots & \phi_{1k} & x_{11} & x_{12} & \cdots & x_{1d} & 1 \\ \phi_{21} & \phi_{22} & \cdots & \phi_{2k} & x_{21} & x_{22} & \cdots & x_{2d} & 1 \\ \phi_{31} & \phi_{32} & \cdots & \phi_{3k} & x_{31} & x_{32} & \cdots & x_{3d} & 1 \\ \vdots & \vdots & \vdots & \vdots & \vdots & \vdots & \vdots & \ddots & \vdots \\ \phi_{m1} & \phi_{m2} & \cdots & \phi_{mk} & x_{m1} & x_{m2} & \cdots & x_{md} & 1 \end{pmatrix} \qquad (2.13)$$

where $\phi_{ij} = exp\left(-\frac{\|\mathbf{x}_i - \mathbf{c}_j\|^2}{2\sigma_j^2}\right)$ values and $\varphi^+ = (\varphi^T \varphi)^{-1} \varphi^T$ denotes the pseudo-inverse of φ. In practice, to avoid possible numerical difficulties due to an ill-conditioned matrix φ, singular value decomposition (SVD) is usually selected to find the weights and independent terms [14].

2.4 Comparison of RBF networks and multilayer perceptrons

Radial-Basis Function (RBF) networks and Multilayer Perceptrons (MLP) are examples of nonlinear layered feedforward networks. They are both universal approximators. It is therefore not surprising to find that there always exists an RBF network capable of accurately mimicking a specified MLP, or vice versa. However, these two networks differ from each other in several important aspects [43]:

1. An RFB network has a single hidden layer, whereas an MLP may have one or more hidden layers.

2. The hidden layer of an RBF network is nonlinear, whereas the output layer is linear. However, the hidden and output layers of an MLP used as a pattern classifier are usually all nonlinear. When the MLP is used to solve nonlinear regression problems, a linear layer for the output is usually the preferred choice.

3. The argument of the activation function of each hidden unit in an RBF network computes the Euclidean norm (distance) between the input vector and the centre of that unit. Meanwhile, the activation function of each hidden unit in an MLP computes the inner product of the input vector and the synaptic weight vector of that unit.

4. MLPs construct global approximations to nonlinear input-output mapping. On the other hand, RBF networks using exponentially decaying localized nonlinearities (e.g., Gaussian functions) construct local approximations to nonlinear input-output mappings.

2.5 Normalized RBFN

By normalizing the equation of the basis function (Eq.2.1), one obtains normalized radial basis functions [45]:

$$\phi_j(\mathbf{x}) = \frac{\phi_j(\| \mathbf{x} - \mathbf{c}_j \|)}{\sum_{i=1}^{M} \phi_j(\| \mathbf{x} - \mathbf{c}_j \|)} \ , \qquad (2.14)$$

where M is the total number of Gaussian kernels. A network obtained using the above form for the basis functions is called a normalized RBFN and has several interesting properties. Firstly this is the natural form of the function obtained in various setting such as noisy data interpolation and regression [14]. Also since the basis function activations are bounded between 0 and 1, they can be interpreted as probability values, especially in classification applications. This model will not be used in this book.

2.6   Summary of RBFN training procedure

The steps in the training process of RBFN including our approach are given below:

Step 1: Choose a number of centroids and train the centroids of the learning data using a stochastic competitive learning according to Eq.2.7.

Step 2: Compute the standard deviation σ_j^c of the learning data in each cluster.

Step 3: Consider a width scaling factor set $Q = [WSF_1, \ldots, WSF_l]$. The widths of the Gaussian kernels are then defined as:

$$\sigma_j = WSF_i \, \sigma_j^c, \qquad 1 \leq j \leq M.$$

Step 4: The weights λ_j and the independent terms a_i and b are computed by a supervised process using singular value decomposition (Eq.2.12).

Step 5: Calculate the error criterion of the validation data.
if $(i == l)$ terminate $else$ go back to step 3: $i = i + 1$.

The optimal WSF_{opt} value will be chosen as the one minimizing the error criterion. If several minima appear, it is recommended to choose the one corresponding to the smallest width scaling factor.

2.7 Conclusion

It is difficult to cover all aspects of learning with local models in one chapter. This is why we have chosen to focus on approximation with Radial Basis Function Networks. Concerning function approximation, we strongly believe in the advantages of RBFN, compared to more conventional multi-layer perceptrons for example. RBFN are easier to use, their performances are less sensitive to local minima and they do not require complex optimization algorithms. We suggested in this chapter a procedure for the computation of the Gaussian function widths based on an exhaustive search belonging option (2) of algorithms quoted in 2.3.2, the purpose being to show the importance of the optimization of Gaussian widths. It was also suggested a one-dimensional searching procedure as a compromise between an exhaustive search on all basis function widths, and a non-optimal a priori choice. In chapter 4, we will show the importance of a proper choice of basis functions widths, and how inadequate value can dramatically influence the approximation performances of the RBFN on several artificial problems.

Chapter 3

Model Selection

3.1 Introduction

In many areas we are faced with the problem of model selection; that is, how complex should we allow our model to be, measured perhaps in terms of the number of free parameters to estimate? If we choose a model that is too complex, then we may be able to model the training data very well (and also any noise on the training data), but it is likely to have poor generalization performance on unknown data, drawn from the same distribution as the training set was drawn from (thus the model overfits the data). Model selection is inherently a part of the process of determining optimum model parameters. In this case, the complexity of the model is a parameter to determine. As a consequence, many model selection procedures are based on optimizing a criterion that penalises a goodness of fit measure by a model complexity measure [85]. In this chapter, we give some general procedures that have been widely used for model selection.

3.2 Separate training, validation and test sets

Before building a model, the samples are often subdivided into "*training*", "*validation*" and "*test*" sets [14]. The distinctions among these subsets are crucial, but the terms *"validation"* and *"test"* sets are often confused in literature.

Training set : The training set is used to train or build a model. For example, in linear regression, the training set is used to fit the linear regression model, i.e. to compute the regression coefficients. In a neural network model, the training set is used to obtain the network weights.

Once a model is built on training data, we need to find the accuracy of the model on unknown data. For this, the model should be used on a set that was not used in the training process. If we were to use the training data itself to compute the accuracy of the model fit, we would get an overly optimistic estimate of the accuracy of the model. This is because the training or model fitting process ensures that the accuracy of the model for the training data is as high as possible – the model is specifically suited to the training data. To get a more realistic estimate of how the model would perform with unknown data, we need to set aside a part of the original data and not use it in the training process. Such sets are known as validation and test sets.

Validation set : The validation set is often used to fine-tune models. For example, we might try out neural network models with various architectures (for example different number of neurons in the hidden layer of RBFN) and test the accuracy of each of the validation sets to choose among the competing architectures.

Test set : When a model is finally chosen, its accuracy on the validation set is still an optimistic estimate of how it would perform with unknown data. This is because the final model has come out as the winner among the competing models based on the fact that its accuracy on the validation set is the highest. Thus, we need to set aside yet another portion of data which is used neither in training nor in validation. This set is known as the test set. The accuracy of the model on the test data gives a realistic estimate of the performance of the model on completely unknown data.

3.3 Model selection

The problem of model selection can be understood as a trade-off between bias and variance [41, 45]. The generalization error can be decomposed into the sum of the squared bias and the variance. A typical trade-off between these two components of generalization error is a function of the model complexity [45]. A too simple model will have a high bias in the sense that, on the average it will differ considerably from the desired one, even though specific instances of the model, obtained by changing the training data, initialization conditions etc..., may hardly differ from one another (figure 3.1 a). On the other hand, a too complex model may have a low bias but will have high variance (figure 3.1 b).

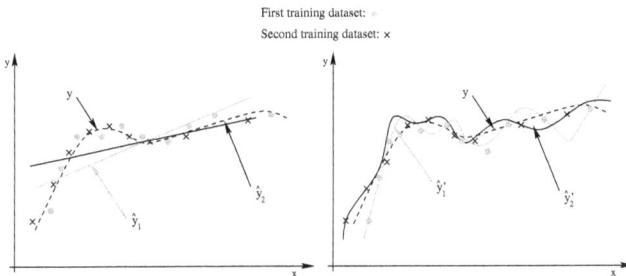

Figure 3.1: Example of two functions: a) High bias and low variance, b) Low bias and high variance [24].

The balancing of bias and variance can be viewed as a problem of finding optimal free parameters. Cross-validation and bootstrap both can be used simply to estimate the generalization error of a given model, or it can be used for model selection by choosing one of several models that has the smallest estimated generalization error. For example, we might use cross-validation to choose the number of neurons in the hidden layer and to choose the width scaling factor WSF (Eq.2.11, page 35) in RBF network.

3.4   Cross-validation

3.4.1   The hold-out method

The hold-out method is the simplest kind of cross-validation. The set is separated randomly into two sets (figure 3.2): a training set used to train a model and a validation set used to estimate the generalization error; i.e. there is no crossing [36, 53, 85]. Usually a training set consists of two-third (2/3) of the data, and a validation set of the remaining third (1/3). The estimated error is computed by:

$$E_{val} = \frac{1}{N_V} \sum_{i=1}^{N_V} (\hat{y}_i - y_i)^2, \tag{3.1}$$

where N_V is the number of samples included in the validation set, y_i are the actual values and $\hat{y}_i$ the values predicted by the model. In this method, the evaluation of the model may depend heavily on which data end up in the training set and which end up in the validation set, and thus the evaluation may be significantly different depending on how the division is made. This will be illustrated in real-world examples in chapter 5.

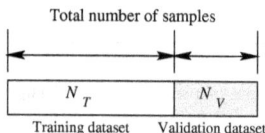

Total number of samples

N_T	N_V
Training dataset	Validation dataset

Figure 3.2: The hold-out method.

3.4.2   Cross-validation (Random subsampling)

Cross-validation also known as random subsampling in the literature, performs K data splits of the entire set, where each split randomly selects a (fixed) number of examples without replacement (figure 3.3). Then for each data split we train or build the model with the training set and then estimate error using the validation set by the following equation:

$$E_{val}^k = \frac{1}{N_V} \sum_{i \in A_V^k} (\hat{y}_i - y_i)^2, \tag{3.2}$$

where N_V is the number of samples included in the validation set A_V^k, y_i are the actual values and $\hat{y}_i$ the values predicted by the model.

The true generalization error estimate is obtained as the average of the separate estimates errors defined as :

$$E_{gen} = \frac{1}{K} \sum_{k=1}^{K} E_{val}^k, \tag{3.3}$$

where E_{val}^k is the estimated error in each experiment and K is the number of repetitions. This estimate is significantly better than the hold-out estimate [36, 78].

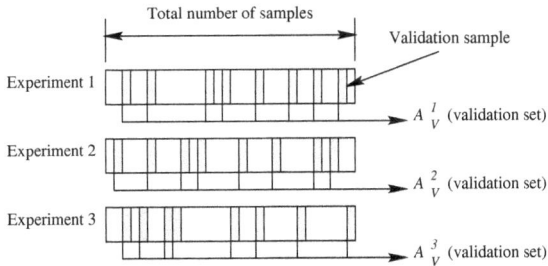

Figure 3.3: Cross-validation.

3.4.3   K-fold cross-validation

In K-fold cross-validation, the set is divided into K subsets of (approximately) equal size. At each iteration, one of the K subsets is used as the validation set and other $K - 1$ subsets are concatenated to form a training set. This procedure is illustrated in figure 3.4, for $K = 4$. For each k value, the estimated error on the validation set is computed by:

$$E_{val}^k = \frac{1}{\frac{N}{K}} \sum_{i \in A_V^k} (\hat{y}_i - y_i)^2, \qquad (3.4)$$

where $\frac{N}{K}$ is the number of samples included in the validation set A_V^k, y_i are the actual values and $\hat{y}_i$ the values predicted by the model.

Then the average error across all K trials is computed:

$$E_{gen} = \frac{1}{K} \sum_{k=1}^{K} E_{val}^k, \qquad (3.5)$$

where E_{val}^k is the estimated error in each experiment and K is the number of repetitions. The advantage of the K-fold cross-validation is that all data in the set are eventually used for both training and validation [26, 40, 41, 78, 85].

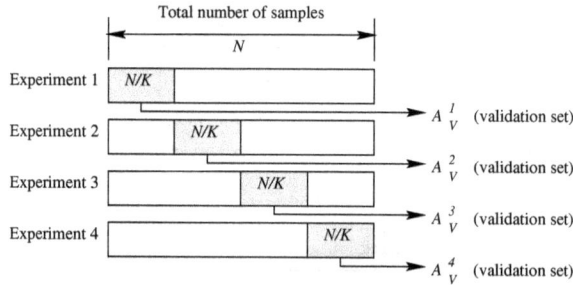

Figure 3.4: K-fold cross-validation.

3.4.4   Leave-one-out cross-validation

Leave-one-out cross-validation is K-fold cross-validation taken to its logical extreme, with K equals N, the number of data in the set. That means that N times, the function approximator is trained on all data except one, and a prediction is made for that data sample (figure 3.5) [40, 41, 49, 78, 85]. The estimated error on the validation set is computed by:

$$E_{val}^k = (\hat{y}_k - y_k)^2, \tag{3.6}$$

where y_k is the actual value and $\hat{y}_k$ the value predicted by the model.

As before the average error is computed and used to evaluate the model,

$$E_{gen} = \frac{1}{N} \sum_{k=1}^{N} E_{val}^k, \tag{3.7}$$

where E_{val}^k is the estimated error in each experiment and N is the number of data included in the set. The evaluation given by leave-one-out cross-validation error is good, but at first pass it seems very expensive to compute. It has often been reserved for problems where relatively small data sizes are available.

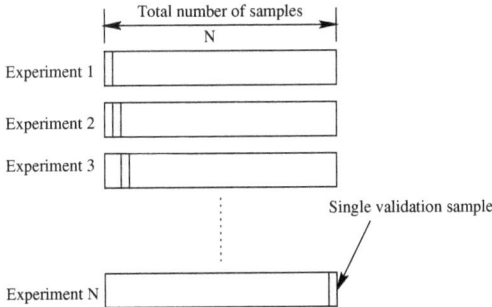

Figure 3.5: Leave-one-out cross-validation.

It should be noted that it is possible to compute E_{gen} without performing the N training in the case of a linear model [62].

3.5   Bootstrap

The bootstrap is a resampling technique with replacement from the original sample [25, 26, 40, 41, 85]. The constructive steps of the bootstrap method, defined in [52] are:

1. From a set X with N samples, one draws randomly N samples with replacement and use this new set X_{learn} for training. The validation set X_{val} is the original learning set.

2. The training of the model g is done using X_{learn}, the errors and $E^k_{learn}(g)$ $E^k_{val}(g)$ obtained with this model are calculated according to the following equations:

$$E^k_{learn}(g) = \frac{1}{N} \sum_{i=1}^{N} (g(x^i_{learn}) - y^i_{learn})^2, \qquad (3.8)$$

with $(x^i_{learn}, y^i_{learn})$ the elements of X_{learn} and $g(x^i_{learn})$ the approximation of y^i_{learn} obtained by model g;

$$E^k_{val}(g) = \frac{1}{N} \sum_{i=1}^{N} (g(x^i_{val}) - y^i_{val})^2, \qquad (3.9)$$

with (x^i_{val}, y^i_{val}) the elements of X_{val} and $g(x^i_{val})$ the approximation of y^i_{val} by model g.

3. The optimism $D^k(g)$, a measure of performance degradation (for the same model) between a learning and a validation set, is computed according to:
$$D^k(g) = E^k_{val}(g) - E^k_{learn}(g). \qquad (3.10)$$

4. Steps 1, 2 and 3 are repeated K times, with K as large as possible. The average optimism $\hat{D}(g)$ is computed by:

$$\hat{D}(g) = \frac{1}{K} \sum_{k=1}^{K} D^k(g). \qquad (3.11)$$

5. Once this average optimism is computed, a new training of the model g is done using the initial set X and $g(x^i)$ the approximation of y^i obtained using the model g.

6. Step 5 is repeated M times, with M as large as possible. For each m $(1 \leq m \leq M)$ the learning error $E_{learn}^m(g)$ is computed over the M repetitions. The average estimate is then computed by:

$$\hat{E}_{learn}(g) = \frac{1}{M} \sum_{m=1}^{M} E_{learn}^m(g). \qquad (3.12)$$

7. Now, we have an estimate of the apparent error and of the optimism, their sum gives an estimate of the generalization error:

$$\hat{E}_{gen}(g) = \hat{E}_{learn}(g) + \hat{D}(g). \qquad (3.13)$$

3.6 Discussion

This chapter has reviewed some general procedures widely used for model selection. Both cross-validation and bootstrapping are methods for estimating the generalization error based on resampling. The resulting estimates of the generalization error are often used for choosing among various models, such as different network architectures.

The estimated error computed by the random subsampling method is better than the hold-out estimate because the hold-out method represents a single train-and-validation experiment. However, the advantage of K-fold cross-validation is that all samples are used for both training and validation and the variance of the resulting estimate is reduced as K is increased. The evaluation given by leave-one-out cross-validation error is good, but its computation cost increases if the number of samples is large. Leave-one-out cross-validation can be less biased but its variance is too high [40, 52].

The bootstrap resampling method may be efficiently used to estimate the generalization error. The main difficulty associated with the bootstrap in real-world applications is the high computation load [77]. In the next chapter we use cross-validation to choose the number of Gaussian kernels in the hidden layer and to choose the width scaling factor in RBF network on several artificial problems.

Chapter 4

RBFN learning: width optimization

4.1 Introduction

In chapter 2, we introduce a procedure to optimize the widths of the Gaussian kernels in order to improve the generalization process. In this chapter, we demonstrate the importance of a proper choice of the widths of the Gaussian kernels, and how inadequate values can strongly influence the approximation performances of the RBFN on several artificial problems. The first section presents a theoretical example with the objective to evaluate the variances of the Gaussian kernels with respect to the dimension of the input space. The next section concerns the analytical examples. It shows the effectiveness of the suggested procedure. For three examples, the proposed procedure is compared to the classical procedures presented in chapter 2. Finally, we finish by conclusion.

4.2 Theoretical example

We consider a simple example, i.e. we try to approximate the unity identity function ($y = 1$) on a d-dimensional hypercube domain $[0, 10]^d$.

It must be mentioned here that this problem is purely theoretical: there is no interest to approximate such a linear (and even constant!)

function by a RBFN. If the RBFN model in Eq. 2.2 (page 29) was used to approximate this function, all weights λ_j multiplying the Gaussian kernels would be equal to zero. In order to reach our goal, i.e. to have insights about the optimal values of the kernel widths, the linear and constant terms were removed from Eq. 2.2 (page 29) in the simulations. Nevertheless, we insist on the fact that the objective of this example is to evaluate the variances of the Gaussian kernels with respect to the dimension of the input space. In order to avoid the consequences of the other parts of the RBFN training algorithm, we chose to work with a constant target function, in order to remove the influence of its variations from our conclusions. Simulations were carried out for different dimensions d, in order to evaluate the influence of the dimension on the results.

For all simulations presented in this section, the density of learning points is uniform in the d-dimensional input space. For this reason, the traditional vector quantization (VQ) first step in the RBFN learning process is skipped; the centroids are attached to the nodes of a square grid in dimension d. The goal of this setting is to eliminate the influence of the VQ results on the simulations. However it is well known that the placement of centroids on the nodes of a square grid is not the ideal result of a vectorial quantization, when $d \geq 2$. For example, it has been demonstrated that in dimension $d = 2$, an ideal vector quantization on a uniform density gives a result in a hive of bee, and not a result in a square grid, as shown it Figure 4.1 [37]. Nevertheless, it can be shown through a simple calculation that the quantization error obtained with the square grid (Fig. 4.1.a) is only about 4% higher than the one obtained with the ideal results (Fig. 4.1.b). As this ideal result is not known in dimensions greater than 2 the assumption is made that the results obtained by placing the centroids on a square grid is a good approximation of those that would be obtained with a true vector quantization.

Once the centroids are placed on a regular grid, the next subsection will show a theoretical way to calculate the optimal width by considering that all the weights are identical in Eq.2.2. Next, in

(a)   (b)

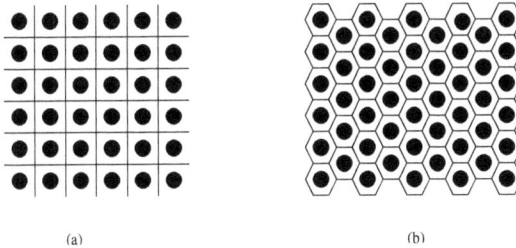

Figure 4.1: (a): Scalar quantization (square grid), (b): vector quantization (hive of bee).

subsection 4.2.2, the optimal width will be estimated by setting all weights λ_j free and calculated according to Eq.2.12.

4.2.1 Theoretical value of the optimal width of the Gaussian kernels

As mentioned above, the centroids are placed on a regular grid, and the function to be approximated is constant ($y = $ constant); therefore it is expected that the weights λ_j in Eq. 2.2 (page 29) will be identical for all centroids. For a theoretical calculation of the optimal *WSF* coefficient, we will make this assumption and further suppose that their values will be equal to 1. Then, we calculate by Eq. 2.2 (page 29) (without linear and constant terms) the theoretical output function of the network, and this for various values of σ_j; again, as the centroids are placed on a regular grid, we will suppose that all σ_j values will be identical. The goal is to find the value of σ_j giving the *flattest* possible output function $\hat{y}$. This one will not be around 1 (there is no reason, since we chose the λ_j equal to 1), but well around another average value m. Taking $\lambda_j = 1$ does not change anything to the problem: if the λ_j were set to $\lambda_j = \frac{1}{m}$, we would have found an output function with an average value of 1, which was the initial problem. Nevertheless, the two problems bring obviously the same conclusions regarding the widths σ_j.

Note that, $\sigma^c = \sigma_j^c$ ($\forall j \in [1,\ldots,M]$) being constant over all clusters, it is equivalent by Eq. 2.11 (page 35) to find a optimal value of σ or a optimal value of *WSF*. In the following subsection, we will estimate optimal values of σ, in order to make possible the comparison with other methods from the literature (subsection 4.2.3).

For each value of σ, to estimate the mean value m we take simply the mean of the output function $\hat{y}$. To quantify the *flatness* of the output function, we calculate its standard deviation (square root of the variance) std_y around the mean value m on validation set. It should be mentioned here that, in order to avoid as much as possible the border effects encountered when using RBFN, we take the mean and the standard deviations of $\hat{y}$ only in the middle of the distribution, i.e. in the $[3.85, 6.05]^d$ compact. For each dimension, the σ giving the flatter function $\hat{y}$ is called **sigma-theo**.

Definition:
Identical Gaussian kernels with unit weights ($\lambda_j = 1$) are summed for various σ values: $\hat{y} = \sum_{j=1}^{M} exp\left(-\frac{\|\mathbf{x}-\mathbf{c}_j\|^2}{2\sigma^2}\right)$, where M is the number of centroids. **Sigma-theo** is the σ value corresponding to the smallest standard deviation of $\hat{y}$.

As an example, figures 4.2(a), 4.2(b), 4.2(c), 4.2(d) and 4.2(e) give the standard deviation std_y of $\hat{y}$ according to σ in 1, 2, 3, 4 and 5 dimensions.

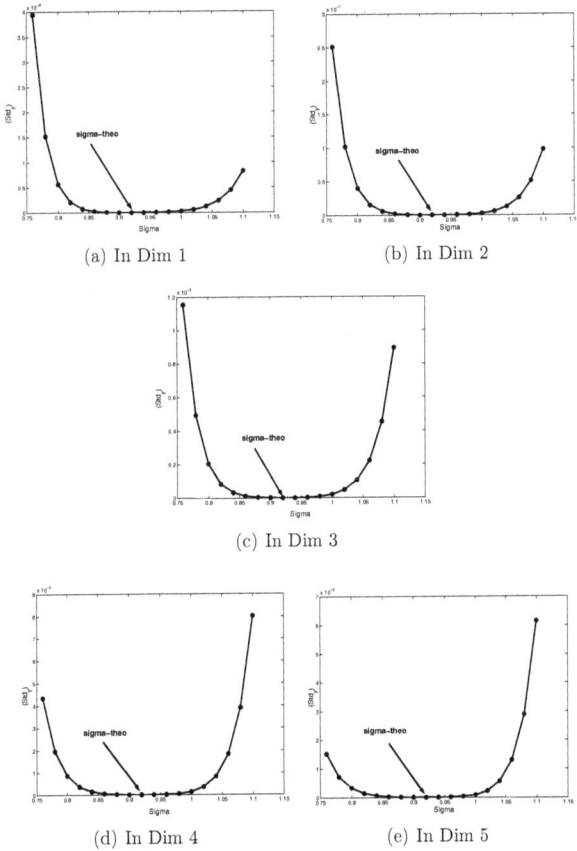

(a) In Dim 1

(b) In Dim 2

(c) In Dim 3

(d) In Dim 4

(e) In Dim 5

Figure 4.2: Standard deviation of the approximation $\hat{y}$, for various space dimensions.

4.2.2   Experimental value of the optimal width of the Gaussian kernels

In this second set of simulation, we still consider centroids placed on a regular grid, but without the assumption that all λ_j weights will be identical. On the contrary, all the weights are set free; we calculate them according to Eq. 2.12 (page 36, without the linear and constant terms) (using Singular Value Decomposition). As in the previous subsection, we repeat the experiment for a large set of possible values σ (identical for all Gaussian kernels), and for several dimensions d of the input space. If the principle of **locality** of the Gaussian kernels is respected and the border effects are neglected, we should expect identical λ_j. In practice, it is not the case, mainly because of the border effects, as shown it figure 4.3(a). As in subsection 4.2.1, we only used the Gaussian kernels in the centre of the distribution (see figure 4.3(b)) in order to decrease the influence of the border effects.

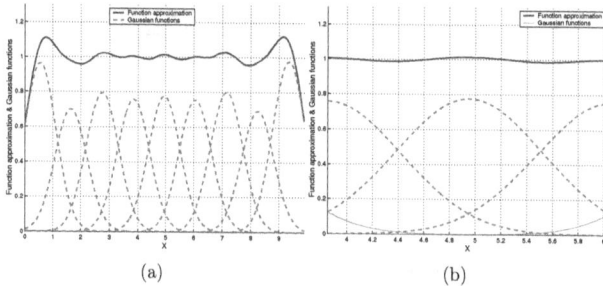

Figure 4.3: (a): With border effects, (b): Without border effects.

After the best-fit function is calculated, the performance of the RBF network is estimated by computing an error criterion. Consider a validation set V, containing N_V data points:

$$V = \{(\mathbf{x}_q, y_q) \in R^d \times R, 1 \leq q \leq N_V : y_q = f(\mathbf{x}_q))\}. \qquad (4.1)$$

The error criterion can be chosen as the mean square error:

$$MSE_V = \frac{1}{N_V} \sum_{q=1}^{N_V} (y_q - \hat{y}_q)^2, \qquad (4.2)$$

where y_q are the desired outputs and $\hat{y}_q$ the predicted values by the model. The minimum of the mean square error (MSE_V) gives now another value of sigma, called **sigma-exp**. It can be verified that the error criterion used for the computation of sigma-exp (MSE_V) is the same criterion as the one used for the computation of sigma-theo (variance).

Definition:
Identical Gaussian kernels are summed for various σ values; $\hat{y} = \sum_{j=1}^{M} \lambda_j \, exp\left(-\frac{\|\mathbf{x}-\mathbf{c}_j\|^2}{2\sigma^2}\right)$, where M is the number of centroids. **Sigma-exp** is the σ value corresponding to the smallest MSE_V.

Figure 4.4(a) gives MSE_V according to σ in dimension 1 and figure 4.4(b) gives the same result in dimension 3. Figures 4.4(a) and 4.4(b)

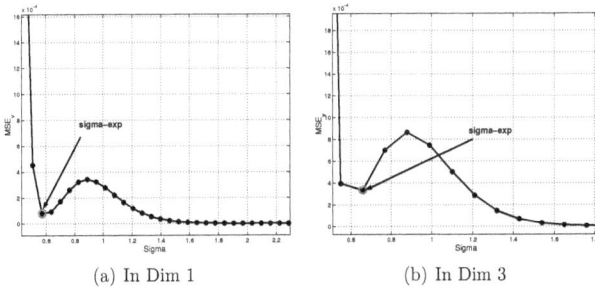

(a) In Dim 1   (b) In Dim 3

Figure 4.4: MSE_v versus σ.

shows the presence of two minimums. The first one corresponds to a local decomposition of the function in a sum of Gaussian kernels; this interpretation is consistent with the classical RBF approach. However, the second one corresponds to a non-local decomposition of the function. As a consequence, the weights λ_j turn out to be enormous in absolute value (positive or negative) in order to compensate for the non-flat slopes. This leads to a greater complexity of the RBFN. In addition, large λ_j dramatically increases numerical instability. The optimal value chosen for σ is the one related to smallest minimum.

4.2.3 Results

The simulations were made on databases of points distributed uniformly in a hypercube of edge lengths equal to 10, in various dimensions d. The number of centroids is chosen equal to 9^d. Other simulations made with a different number of centroids gave similar results. The number of training and validation points is chosen sufficiently large to avoid (as much as possible) falling into the difficulties due to the empty space phenomenon (between 1000 and 50000 training points according to the dimension of the input space). These restrictions have limited the simulations to a dimension of input space equal to 5.

Table 4.1 gives the results. In order to obtain values independent from the number of centroids, the "Width Scaling Factors (Eq. 2.11)" WSF-theo and WSF-exp are defined, as being the ration between sigma-theo and sigma-cluster on one hand, and sigma-exp and sigma-cluster on the other hand respectively. Indeed it is more appropriate to compare results on the scale-independent *WSF* coefficient instead of the values of σ for two reasons:

- most results in the literature are based on the sigma-cluster value, making the use of *WSF* easier for comparisons;

- the *WSF* values are independent from the number of centroids, while those of σ are not.

Dim	sigma-cluster	sigma-theo	sigma-exp	WSF-theo	WSF-exp
1	0.3175	0.92	0.5715	2.897	1.8
2	0.4490	0.92	0.5837	2.049	1.30
3	0.5499	0.92	0.5506	1.673	1.00
4	0.6350	0.92	0.5676	1.448	0.89
5	0.7099	0.92	0.5471	1.296	0.77

Table 4.1: WSF-theo and WSF-exp according to the dimension of the input space.

Several comments result from table 4.1:

- The sigma-cluster is proportional to the square root of dimension, as shown by simple analytical calculation:

$$\sigma_{cluster} = \sqrt{d}\frac{a}{2\sqrt{3}}, \tag{4.3}$$

  where a is the length of an edge of the hypercube corresponding to the Voronoi zone of a centroid. In the simulations, the 9^d centroids are placed a priori at the positions $0.55 + k * 1.1$ (with $1 \leq k \leq 9$) measured on each axis of input space; a is thus equal to 1.1.

- We notice that the sigma-theo does not depend on the dimension of the input space. Therefore, WSF-theo is inversely proportional to the square root of the dimension of the input space (see Eq. 4.3).

- We also notice that the sigma-exp values are systematically lower, about 30 to 35%, than the sigma-theo values. This is due to the increased freedom given to the network coefficients by allowing weight variations rather than fixing them.

We also compared our method (calculation of **sigma-exp**) to the three approaches of Moody et al. [60], S. Haykin [43] and A. Saha et al. [70], quoted in chapter 2. Figures 4.5, 4.6 and 4.7 illustrate the mean square error obtained according to sigma for different dimension d ($1 \leq d \leq 3$) with the various calculation methods of σ. We notice here that whatever is the dimension of the input space, we always find two minima. Figures 4.8(a), 4.8(b), 4.9(a) and 4.9(b) clearly show that the choice of the sigma value has great influence on the local character of the decomposition in a sum of Gaussian kernels of the function (in dimension 1) to approximate.

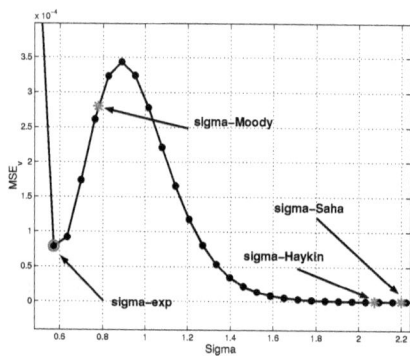

Figure 4.5: MSE_v according to sigma in Dim 1 with the various calculation methods of σ.

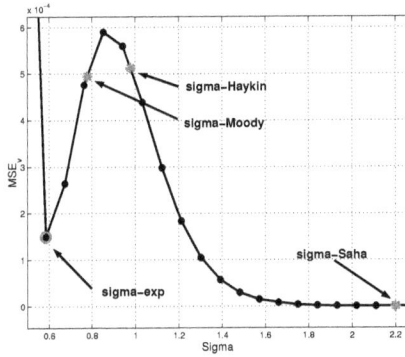

Figure 4.6: MSE_v according to sigma in Dim 2 with the various calculation methods of σ.

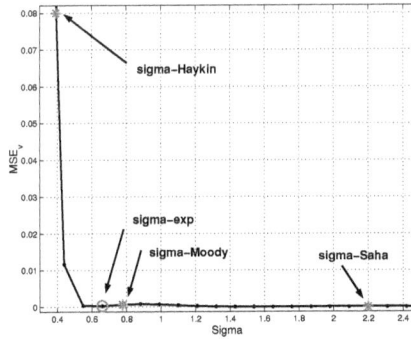

Figure 4.7: MSE_v according to sigma in Dim 3 with the various calculation methods of σ.

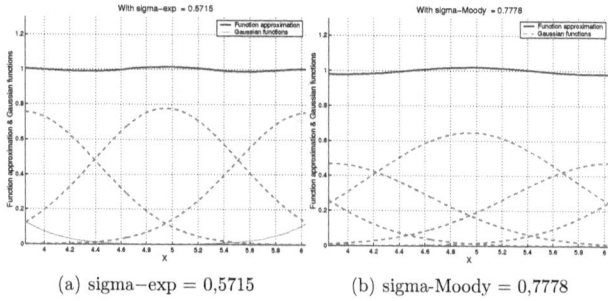

(a) sigma−exp = 0,5715   (b) sigma-Moody = 0,7778

Figure 4.8: Decomposition in a sum of Gaussian kernels of $y = 1$.

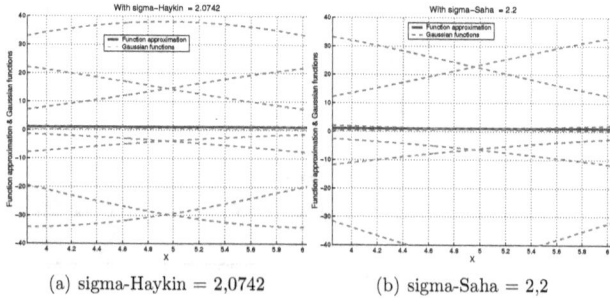

(a) sigma-Haykin = 2,0742   (b) sigma-Saha = 2,2

Figure 4.9: Decomposition in a sum of Gaussian kernels of $y = 1$.

4.3 Analytical example

It is shown in this section the importance of optimizing the width scaling factor WSF in the case of the approximation of three analytical functions, in addition to a comparison between our method and the methods quoted in the chapter 2 (section 2.3.2). We should mention here that the RBF network model Eq. 2.2 (page 29) is used without the linear and constant terms to approximate the three analytical functions.

4.3.1 Approximation results

Firstly, we consider N data selected randomly according to a Dim 1 sine wave [6]:

$$x \in [0,\ 1] : y_1 = \sin(12\ x). \tag{4.4}$$

In this example, we used cross-validation (random subsampling with $K = 20$ experiments) to choose the optimal width scaling factor WSF (Eq.2.11). For each data split we build the model with the training set (400 samples) and then estimate the error with the validation set (100 samples). The true error estimate is obtained as the average of the separate estimated errors defined by Eq. 3.3. The function and its corresponding approximation are plotted in figure 4.10. In figure 4.11, we have plotted the MSE_V as function of WSF. One can notice that in some cases a second minimum appears for higher WSF. Though, the second minimum is rather fortuitous than systematic, as the average curve confirms. The optimal value for WSF is thus taken as the first minimum, i.e. approximately $WSF = 2.35$. It should be noted here, that to determine the optimal number of centers in the hidden layer, we tested RBF networks with different number of centers. The optimal number of centers to approximate y_1 is 5.

Secondly, we consider the following function [6]:

$$x \in [-4,\ 4] : y_2 = 1 + (x + 2x^2)\sin(-x^2). \tag{4.5}$$

In this example, cross-validation is used as well (random subsampling with $k = 50$ experiments) to separate randomly the whole set in two sets. The first set is a training set (4000 samples) used to train a model

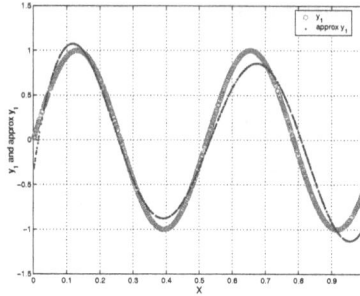

Figure 4.10: 1D sine wave and its approximation by a RBF network.

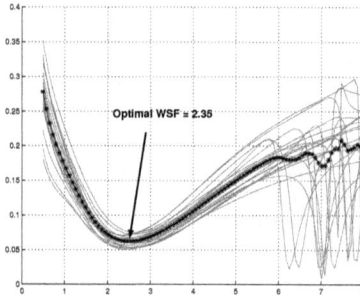

Figure 4.11: $MSE_V(\text{-})$ and mean curve (thick line) according to WSF for 1D sine wave.

and the second set is a validation set (1000 samples) used to estimate the generalization error. The function and the RBFN approximation are illustrated in figure 4.12, while its MSE_V is plotted in figure 4.13. Here again we observe two minima. However, both are systematic, as the average curve demonstrates. Nevertheless, the two minima are of a different type. The first minimum corresponds to a *local* decomposition of the function in a sum of Gaussian functions (figure 4.16(a)). The second one, on the contrary, corresponds to a *non-local* decomposition of the function (figure 4.16(b)). The optimal width factor scaling is the one related with the smaller WSF ($WSF = 1.8$, see figure 4.13). In this example to determine the optimal number of centers in the hidden layer, we tested RBF networks with various number of centers. The optimal number of centers used in the hidden layer to approximate y_2 is 20.

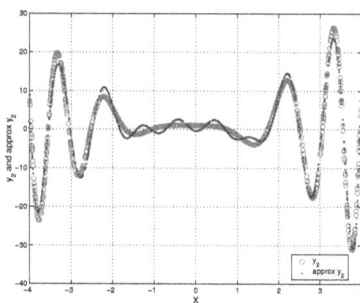

Figure 4.12: Function y_2 and its approximation by a RBF network.

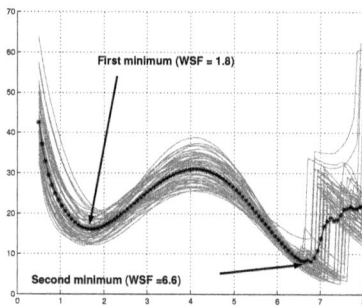

Figure 4.13: $MSE_V(\text{-})$ and mean curve (thick line) according to WSF for y_2.

Thirdly, we consider N data selected randomly according to the following cosinus wave in dimension 2:

$$x \in [0, 3] : y_3 = 5\cos(x_1 \cdot x_2 - 2x_1) - 7. \qquad (4.6)$$

In this example, also we used the cross-validation method (random sub-sampling with K=50 experiments) to choose the optimal width scaling factor WSF (Eq.2.11). For each k value, we draw randomly two-third $(2/3)$ from the data $(N = 1000)$ to form the training set (666 samples). The remaining third part forms the validation set (334 samples). The true error estimate is obtained as the average of the separate estimates errors defined by Eq. 3.3. Figure 4.14 shows the estimation of the mean square error on validation set (MSE_V) by the cross-validation method. The optimal value of the width scaling factor (WSF) is taken as the minimum of the average curve, i.e. approximately $WSF = 3.48$ (see figure 4.14). Also in this example, we tested RBF networks with different number of centers in the hidden layer. The optimal number of centers used in the hidden layer to approximate y_3 is 8.

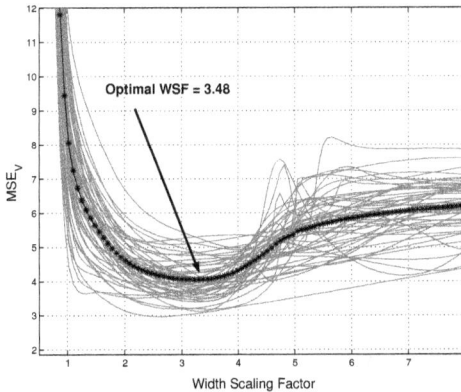

Figure 4.14: MSE_V(-) and mean curve (thick line) according to WSF for y_3.

4.3.2 Comparison

We have compared our approach to the classical approaches given by Moody et al. [60] and S. Haykin [43], quoted in chapter 2. Figures 4.15(a), 4.15(b) and 4.15(c) show the decomposition in a sum of Gaussian kernels of y_1 with our approach and with the classical methods given by Moody et al. and S. Haykin respectively. Figures 4.16(a), 4.16(b), 4.17(a) and 4.17(b) show the decomposition in a sum of Gaussian kernels of y_2 with the same methods. These figures in both examples clearly show that the choice of the widths of the Gaussian kernels in RBF networks has a **significant influence** on the local character of the decomposition in a sum of Gaussian kernels of the function to approximate.

	Method	MSE_V	**Locality**
	Moody and Darken	0,0844	High
y_1	S. Haykin	0.1334	Medium
	Our approach	**0.0625**	High
	Moody and Darken	24.7994	High
y_2	S. Haykin	5.3929	Low
	Our approach	**16.1266**	High
	Moody and Darken	6.1454	Medium
y_3	S. Haykin	6.2364	Medium
	Our approach	**4.0506**	High

Table 4.2: Comparison of the performances of our approach and classical approaches.

Table 4.2 gives the mean square errors obtained for the three examples by using the various methods for the calculation of the widths of the Gaussian kernels. In the three examples our approach exhibits the best compromise between accuracy and complexity. Indeed, we obtain small mean square errors combined with greater locality and small λ_j.

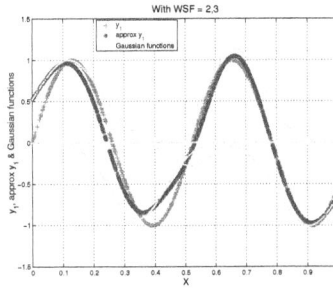

(a) with our approach

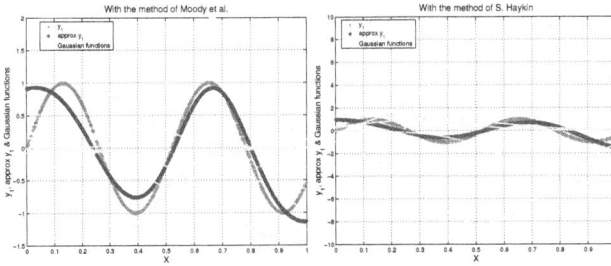

(b) with the method of Moody et al.   (c) with the method of S. Haykin

Figure 4.15: Decomposition in a sum of Gaussian kernels of y_1.

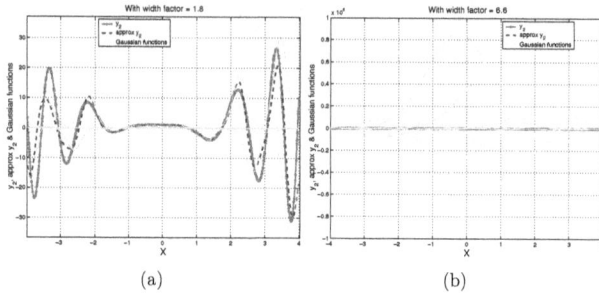

(a)   (b)

Figure 4.16: (a): Local decomposition of y_2 with first minimum, (b): Non-Local decomposition of y_2 with second minimum.

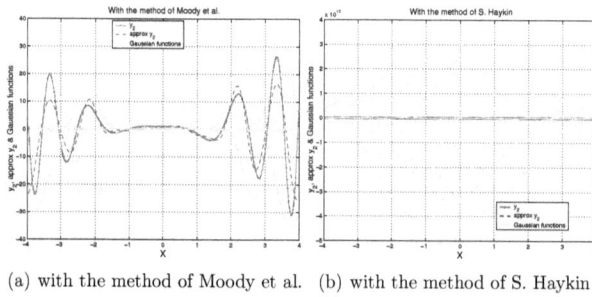

(a) with the method of Moody et al.   (b) with the method of S. Haykin

Figure 4.17: Decomposition in a sum of Gaussian kernels of y_2.

4.4   Conclusion

We demonstrated in this chapter the importance of the choice of the widths of the Gaussian kernels. In many situations, a bad choice can lead to an approximation error definitely higher than the optimum, sometimes by several orders of magnitude. Subsequently we showed using simulations, that a classic choice (taking the width of the Gaussian kernels equal to the standard deviation of the points in a cluster) is certainly not optimal. For example, in a theoretical example in dimension 1, it appears that the width should be twice this value. It was also shown that the dimension of the data space has an important influence on the choice of σ. In particular, the multiplicative correction that must be applied to the standard deviation of points in a cluster is inversely proportional to the square root of the dimension of the input space. Finally, in the three analytical examples our approach exhibited the best compromise between accuracy and complexity. Indeed, we obtained small mean square errors combined with greater locality and small multiplicative weights (λ_j).

The results also showed the need for a greater attention to be given to the optimization of the widths of the Gaussian kernels in RBF networks, and to the development of methods that allow fixing these widths according to the problem without using exhaustive search.

Part II

Variable selection

Chapter 5

Spectrophotometric variable selection

5.1 Introduction

In recent years, qualitative and quantitative applications of infrared spectroscopy in various chemical fields including the pharmaceutical, food and textile industries have grown dramatically [16]. The chemical analysis by spectrophotometry is based on the fast acquisition of a great number of spectral data (several hundreds, even several thousands).

According to the Geladi [33], the chemometrics is a new chemical branch which uses the theory and the models developed in statistics, mathematics and computer sciences to extract useful and substantial information from chemical measurements. In analytical applications, the most frequently encountered case corresponds to the prediction of a quantitative variable, such as the concentration of a component present in the studied product.

From a chemometric point of view, the spectral data have remarkable characteristics, which requires specific methods for their treatment. The matrix **X** of the data may comprise **more variables (spectral data) than observations (spectra)**. This case is rather less encountered in other applications of statistics. Before applying a chemometric method it is necessary to create a collection of spectra to which the study will relate and which will be used to establish a linear or non-linear model.

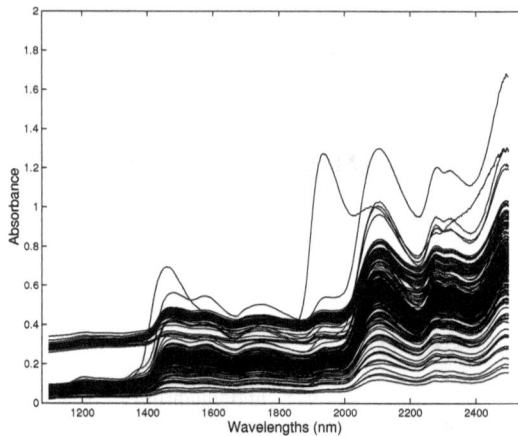

Figure 5.1: Example of near-infrared reflectance spectra of 218 orange juice.

Figure 5.1 shows a collection of 218 near-infrared reflectance (R) spectra in the wavelength range from 1100 to 2500 nm with a step of 2 nm. These data are gathered in a matrix $\mathbf{X}$, including in this example 218 lines and 700 columns. Figure 5.2 shows the graph of the absorbance $(\log \frac{1}{R})$ at two wavelengths for the 218 orange juice spectra. For two consecutive wavelengths (variables) in the spectrum (2298 and 2300 nm), the points representing the individuals are placed almost exactly on a line. For a distance from 1000 nm between the wavelengths (1422 and 2422 nm) the points are not aligned.

We conceive that certain columns (variables) of the matrix $\mathbf{X}$ can be practically represented as a linear combination of other columns. This

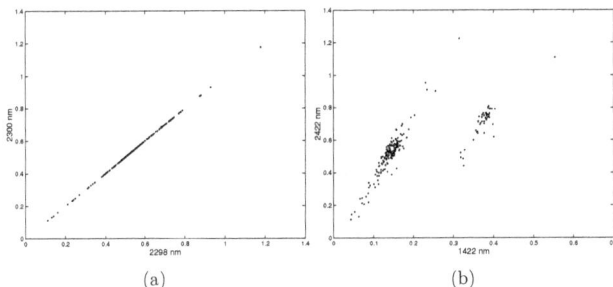

(a)   (b)

Figure 5.2: (a) The two consecutive wavelengths 2298 and 2300 nm are very collinear, (b) The two wavelengths 1422 and 2422 nm are not collinear.

situation is called **collinearity** and is the source of many problems in the direct application of many statistical methods, such as the multiple linear regression (MLR) [5, 13, 27, 31, 54, 62]. Studies have shown that if collinearity is present among the variables, the prediction results can get poor (see for example [27, 79]). This limitation has promoted other alternative linear methods to offset the problems engendered by the strong redundancy between the spectral data. Several alternatives that are able to adapt to this collinearity were developed: Stepwise multiple linear regression (SMLR) [13, 23, 38, 55, 62], Principal component regression (PCR) [31, 34, 40, 54, 55] and Partial least square regression (PLSR) [13, 34, 40, 44, 54, 55, 80, 88], etc.

Often multivariate calibration techniques for near-infrared spectroscopy assume a **linear relationship** between dependent variables (e.g. concentration of target analyte) and independent variables (e.g. absorbances at different wavelengths); while a strictly linear relationship rarely exists, both PCR and PLSR provide correct results as long as the deviation from linearity is not too large. However, with strong non-linearity both calibration techniques lead to substantial errors so an alternative chemometric tool must be used instead. Non-linearity in near-infrared spectroscopy arises from various factors, namely

[12, 16, 20, 22, 57, 59]:

- deviations from the Beer-Lambert's law, which are typical of highly absorbing samples;

- non-linear detector responses;

- drifts in the light source;

- interactions between analytes;

- non-linearity in diffuse reflectance and transmittance data .

This type of non-linearity can be modeled by using non-linear methods, for example artificial neural networks.

To summarize, spectral data obtained from spectrophotometers have remarkable characteristics:

- more variables (spectral data) than observations (spectra);

- collinearity;

- non-linear relationship between dependent variables and independent ones.

Therefore, there is a need to select variables among the candidates, in order to build a still suitable model with only few variables. In this chapter firstly we present a new incremental method (step by step) for the selection of spectral variables using linear and non-linear regression (the latter being implemented by Radial basis functions networks). The approach that was proposed in chapter 2 will be used to perform the learning of RBF networks. Secondly we suggest a way to improve this method by a judicious choice of the first spectral data, which has an important influence on the final performances of the modeling. The choice of the first variable is based on the measurement of the mutual information between the spectral data (independent variables) and the concentration (dependent variable).

The chapter is organized in 7 main sections. Beer-Lambert's law is presented in section 5.2. Section 5.3 presents the usual techniques for spectrophotometric variables selection in the context of linear regression methods: SMLR (Stepwise Multiple Linear Regression), PCR (Principal Component Regression) and PLSR (Partial Least Square Regression). Non-linear regression models (RBF networks) for spectrophotometric variable selection, through an incremental procedure based on a validation criterion are presented in section 5.4. The predictive ability of both linear and non-linear models for spectrophotometric variables selection is compared on four real-life datasets in terms of normalized mean square error on validation set. The result of this predictive ability comparison is presented in section 5.5. In section 5.6 four improvements of the previous method are presented. The use of the cross-validation method to select the relevant variables in the wine dataset is presented in subsection 5.6.1. In the following subsection the graphical detection of outliers concerning the spectrophotometric data is described. The data distribution method between training and validation sets is presented in subsection 5.6.3. An improved method for spectrophotometric variable selection using non-linear models is introduced in subsection 5.6.4. Some concluding remarks are given in the final section.

5.2   Beer-Lambert's law

Spectrophotometric analysis relies on the interaction of electromagnetic radiation (light) with the matter of interest. Strictly speaking, every compound has a distinct absorption spectrum, which allows its identification, in many cases, in the presence of other compounds. In addition to the identification of a compound, it is also possible to determine quantitatively the concentration of that compound. The relationship between absorbance and concentration is given by the Lambert-Beer's law and is written mathematically as:

$$A = \epsilon\, C\, x, \tag{5.1}$$

where A is the absorbance, ϵ is the molar absorptivity constant, x is the pathlength over which the light interacts with the sample in cm and C is the concentration.

In a more practical sense, the absorbance is defined as the negative logarithm of the transmittance. This is given mathematically as:

$$A = -\log T = -\log \frac{I}{I_0},\tag{5.2}$$

where I is the intensity of the light beam on the sample, I_0 is the intensity of the light beam after having passed the sample and T is transmittance.

Achá et al. [1] showed that, in infrared transmittance the relationship between the absorbance and the concentration $\log \frac{1}{T}$ is described by the Beer-Lambert's law, i.e. this relation is linear. However, that law should not be applied to near-infrared diffuse reflectance $\log \frac{1}{R}$, because of the light scattering according to Alfaro et al. [3] and the light pathlength shortening according to Meurens [57], as shown in figure 5.3.

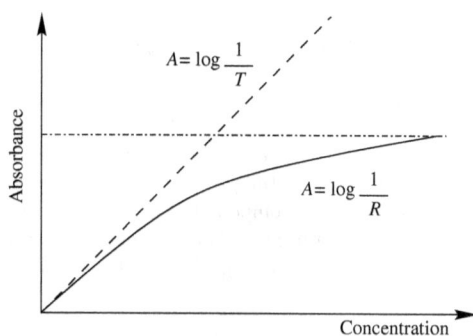

Figure 5.3: Absorbance according concentration.

5.3 Spectrophotometric variable selection by linear models: state of the art

A model is a relationship $y = f(\mathbf{x})$ between two groups of variables, often called dependent variable y and independent variables $\mathbf{x}$. By independent variables, we mean spectral data. Depending on the research discipline, the spectral data are either regarded as independent variables (e.g. in chemometrics) or as dependent variables (e.g in spectroscopy); this last definition comes from the fact that the shape of the spectrum depends on the analyte concentration present in a given studied product.

Chemical analysis usually consists of two steps [34]. First, a model or a family of models is chosen, and their parameters are chosen according to a set of data. The dataset used for this step is called a calibration or training set (see definition of the training set in section 3.2). The model parameters are called regression coefficients. The second step is the one in which more observations of the independent variables are used. These are used together with the regression coefficients to predict values of the dependent variables. This is the prediction or validation step. The set used in this step is the prediction or validation set (see definition of the validation set in section 3.2).

5.3.1 Multiple linear regression (MLR)

The question is to predict a dependent variable y from independent variables $x_1, x_2, \ldots, x_n$ which are spectral data (variables) measured on various wavelengths or wavenumbers. The multiple linear regression (MLR) model in its matrix form is [23, 34, 38, 55, 62]:

$$\mathbf{y} = \mathbf{X}\mathbf{b} + \mathbf{e} \qquad (5.3)$$

where $\mathbf{y}$ is a $(m \times 1)$ vector of measured responses (dependent variables), $\mathbf{X}$ is a $(m \times n + 1)$ matrix of measured spectra (independent variables) augmented with a column of ones, $\mathbf{b}$ is a $(n + 1 \times 1)$ vector of regression coefficients and $\mathbf{e}$ is a vector of residuals.

The estimation of the unknown parameters constituting the vector

b is realised by minimizing cost function, for example residual sum of squares:

$$SS_{Res} = \sum_{i=1}^{m} (\hat{y}_i - y_i)^2. \tag{5.4}$$

There are three possible ways to resolve the equation 5.3:

1. When the number of samples (observations) and variables are equal $(m = n + 1)$ then there is a unique solution for **b**:

$$\mathbf{X}^{-1}\mathbf{y} = \mathbf{X}^{-1}\mathbf{X}\mathbf{b}, \tag{5.5}$$

$$\mathbf{b} = \mathbf{X}^{-1}\mathbf{y}. \tag{5.6}$$

2. If the number of samples is greater than the number of variables $(m > n + 1)$ then a least squares solution for **b** is obtained by forming the generalized inverse of **X**:

$$\mathbf{X}^T\mathbf{y} = \mathbf{X}^T\mathbf{X}\mathbf{b}, \tag{5.7}$$

$$(\mathbf{X}^T\mathbf{X})^{-1}\mathbf{X}^T\mathbf{y} = (\mathbf{X}^T\mathbf{X})^{-1}\mathbf{X}^T\mathbf{X}\mathbf{b}, \tag{5.8}$$

$$\mathbf{b} = (\mathbf{X}^T\mathbf{X})^{-1}\mathbf{X}^T\mathbf{y}. \tag{5.9}$$

Equation 5.9 gives a hint towards the most frequent problem in MLR: the inverse of $(\mathbf{X}^T\mathbf{X})$ may not exist [31, 34, 62].

3. If the number of samples is lower than the number of variables $(m < n+1)$ then there are infinite number of solutions for **b**. There exist many techniques to find one specific solution. The minimum norm solution to this least square problem is given by [35]:

$$\mathbf{b} = \mathbf{X}^T(\mathbf{X}\mathbf{X}^T)^{-1}\mathbf{y}. \tag{5.10}$$

Note that, often the matrix **X** comprises more variables than observations, then collinearity is guaranteed to occur.

We have to remember at this stage that only non-singular (determinant is non-zero) square matrices have inverses. The technique shown above in the second case is to obtain a square matrix $(n+1 \times n+1)$ by multiplying **X** by its transpose $\mathbf{X}^T$. However the inverse of this product matrix can exist only when the resulting matrix is non-singular [31, 34, 62]. In the third case, the matrix **X** comprises more variables than samples, then collinearity is guaranteed to occur. The solution of this problem is to delete some variables using procedures of variable selection [23].

5.3.1.1   The coefficient of multiple determination

A manner to evaluate the relevance of linear model consists in measuring the variation of y explained by the model. The regression equation is estimated such that the total sum-of-squares can be partitioned into components due to regression and residuals [13, 55]:

$$\sum_{i=1}^{m}(y_i - \bar{y})^2 = \sum_{i=1}^{m}(\hat{y}_i - \bar{y})^2 + \sum_{i=1}^{m}(\hat{y}_i - y_i)^2$$

$$SS_{Tot} \quad = \quad SS_{Reg} + \quad SS_{Res} \tag{5.11}$$

The quality of adjustment of the multiple linear regression model can be determined using the coefficient of multiple determination R^2 (called also square of the coefficient of multiple correlation), defined by:

$$R^2 = \frac{SS_{Reg}}{SS_{Tot}} = 1 - \frac{SS_{Res}}{SS_{Tot}} \tag{5.12}$$

If the regression is perfect, all residuals are zero, SS_{Res} is zero, and $R^2 = 1$. If there is no linear relationship between the dependent and independent variables, then R^2 is equal to 0.

5.3.1.2   Collinearity

Collinearity is present when the columns of $\mathbf{X}$ are approximately or exactly linearly dependent. In the case of exact linear dependency, $(\mathbf{X}^T\mathbf{X})^{-1}$ is not defined and consequently the vector $\mathbf{b}$ of regression coefficients can not be expressed by the equation 5.9. If the linear dependency is approximate, at least one of the diagonal elements in the covariance matrix, $(\mathbf{X}^T\mathbf{X})^{-1}$, will be large. This leads to unstable estimates of the regression coefficients which may be unreasonably large (in absolute value). High correlation of $\mathbf{x}$ variables will easily lead to unreliable predictions. Therefore, it is important to be able to detect whether $\mathbf{X}$ is collinear or not, prior to regression analysis.

A related indication of collinearity is the *variance inflation factor* (VIF) [55, 62]:

$$VIF_i = \frac{1}{(1 - R_i^2)}, \tag{5.13}$$

where R_i^2 is the coefficient of multiple determination when $\mathbf{x}_i$ (the ith variable in $\mathbf{X}$ considered here as the dependent variable) is regressed on the remaining variables. When the columns of $\mathbf{X}$ are close to linear dependence, R_i^2 will be close to unity and VIF_i will be large. A VIF greater than 5 is generally considered large and is an indication that the corresponding coefficient is poorly estimated [55].

Note that, when the measurements are made in infrared spectroscopy, often the matrix $\mathbf{X}$ comprises more variables than samples, then collinearity is guaranteed to occur. In this situation, a form of variable selection is required.

The problem of variable selection can be defined as follows: given a set of candidate variables, select a subset that performs best (according to some criterion) in a prediction system. More specifically, let $\mathbf{X}$ be the original matrix of spectral data, containing n different variables (columns of $\mathbf{X}$) and m observations (row of $\mathbf{X}$). The objective is to find a subset of the columns of $\mathbf{X}$, $Z \subseteq \mathbf{X}$ containing d variables representing the best model [27, 58].

5.3.2 Stepwise multiple linear regression (SMLR)

It is possible to carry out a test of assumptions in order to judge if the variable $\mathbf{x}$ explains a significant part of the variability of y. The general step of test of assumptions is consigned in a table (Table 5.1) known as counts of the analysis of variance (ANOVA) associated with the model of multiple linear regression. Analysis of variance (ANOVA) allows us to verify whether the predictor variables can explain a significant amount of the variance in the response variable. In this table MS are mean squared terms, SS are sum of squares terms, p is the number of regression coefficients, and m is the number of observations (samples). F is the ratio of the mean squared terms; it estimates the statistical significance of the regression equation.

Of course with a larger number of $\mathbf{x}$ variables the comparison of all possible obtained models (results) requires a lot of computation time. A stepwise multiple linear regression procedure is generally preferred over

Source of variation	SS	Degrees of freedom (df)	MS	F
Regression	SS_{Reg}	$p-1$	$MS_{Reg} = \frac{SS_{Reg}}{p-1}$	$\frac{MS_{Reg}}{MS_{Res}}$
Residual	SS_{Res}	$m-p$	$MS_{Res} = \frac{SS_{Res}}{m-p}$	
Total	SS_{Tot}	$m-1$		

Table 5.1: Analysis of variance table for multiple linear regression.

this brute force approach [13, 55, 62].

To observe the changes in the regression sum of squares, we consider, for example, the following models:

Model 1: $\hat{y} = b_0 + b_1 x_1$ $SS_{Reg}(1)$
Model 2: $\hat{y} = b'_0 + b'_1 x_1 + b'_2 x_2$ $SS_{Reg}(2)$

If $SS_{Reg}(1)$ and $SS_{Reg}(2)$ are the regression sum of squares for these models then $SS_{Reg}(2) - SS_{Reg}(1)$ represents the increase of the regression sum of squares due to the inclusion of x_2 in the model. It is called $SS_{Reg}(x_2 \mid x_1)$, the sum of squares due to x_2 given x_1 is already in the model. Since $SS_{Reg}(1)$ and $SS_{Reg}(2)$ have 1 and 2 degrees of freedom, respectively, there is 1 degree of freedom associated with $SS_{reg}(x_2 \mid x_1)$. The corresponding mean square, $MS(x_2 \mid x_1) = SS_{Reg}(x_2 \mid x_1)$, is compared with $MS_{Res}(2)$, the residual mean square for the more complex model, by means of an F test [55, 62]:

$$F = \frac{MS(x_2 \mid x_1)}{MS_{Res}(2)} = \frac{SS_{Reg}(2) - SS_{Reg}(1)}{MS_{Res}(2)}. \tag{5.14}$$

This F test is called a partial F-test and is important for the selection of relevant variables in the stepwise multiple linear regression (SMLR) described below. In the forward step of SMLR, F test of Eq. 5.14 is called F-to-enter and is defined as the partial F-test performed on a variable which is not yet in the model. In the backward step of SMLR, F test of Eq. 5.14 is called F-to-remove and is defined as the partial

F-test performed on a variable already in the model as though it was added last to the model.

The stepwise multiple linear regression (SMLR) procedure works as follows. The spectral variables are selected among the n available variables by respecting a criterion of optimization such as the test of assumption based on Fisher's law. The selection starts with the variable that has the largest correlation with the dependent variable. At each step the F-to-enter values for all variables not yet in the model are checked and the highest significant F-to-enter value is compared to Fisher table ($F_{\alpha, df1, df2}$, where $\alpha = 5\%$ or 10%, *df1* and *df2* are the degrees of freedom for the regression mean square and residual mean square from the ANOVA table). If the variable with F-to-enter value is higher than $F_{\alpha, df1, df2}$, then this variable is selected. Moreover, after each step the F-to-remove values for all variables already in the model are tested and the smallest significant F-to-remove value is compared to Fisher table ($F_{\alpha, df1, df2}$). If this F-to-remove value is less than $F_{\alpha, df1, df2}$, then this variable is removed. The procedure is continued until no more variable fulfils the criterion to be selected or to be removed [13, 55].

5.3.2.1 Example

The following SMLR procedure example is taken from Massart et al.'s book [55]. We note that, the matrix **X** consists of 17 samples in dimension 3. For this example, we used the adapted stack loss dataset given in appendix A, page 151.

The necessary information for the calculations is summarized in table 5.2.

variables	SS_{Reg}	df	SS_{Res}	df	$R^2\,\%$
x_1	775.482	1	40.753	15	95.01
x_2	567.032	1	249.203	15	69.47
x_3	134.799	1	681.436	15	16.51
x_1 and x_2	739.975	2	22.260	14	97.25
x_1 and x_3	776.845	2	39.390	14	95.17
x_2 and x_3	576.279	2	239.956	14	70.60
x_1, x_2 and x_3	795.834	3	20.401	13	97.50

Table 5.2: Stack loss data regression sum of squares, residuals sum of squares and % variation explained for different regression equations.

	x_1	x_2	x_3	y
x_1	1.000	0.7554	0.454	0.975
x_2		1.000	0.369	0.833
x_3			1.000	0.406
y				1.000

Table 5.3: Correlation matrix for the stack loss data.

Step 1: The selection starts with the variable that has the largest correlation with the dependent variable. From the correlation matrix (Table 5.3) it follows that the dependent variable y is most correlated with x_1; therefore, this variable is the first to enter the regression equation.

Step 2: For the variables not yet in the model the following F-to-enter values are calculated:

$$x_2: \quad \text{F-to-enter} = \frac{SS_{Reg}(x_1,x_2) - SS_{Reg}(x_1)}{MS_{Res}(x_2;x_1)}$$

$$x_2: \quad \text{F-to-enter} = \frac{(793.975 - 775.842)}{\frac{22.260}{14}} = 11.63 > F_{0.05;1,14}(= 4.60)$$

x_3 : F-to-enter $= \frac{SS_{Reg}(x_1,x_3)-SS_{Reg}(x_1)}{MS_{Res}(x_3,x_1)}$

x_3 : F-to-enter $= \frac{(776.845-775.842)}{\frac{39.390}{14}} = 0.48 < F_{0.05;1,14}(= 4.60)$

It follows that x_2 has the highest F-to-enter value. Since it contributes significantly to the regression, x_2 is added to the model.

Step 3: For the variables in the regression equation (x_1 and x_2) the following F-to-remove values are calculated:

x_1 : F-to-remove $= \frac{SS_{Reg}(x_1,x_2)-SS_{Reg}(x_2)}{MS_{Res}(x_2,x_1)}$

x_1 : F-to-remove $= \frac{(793.975-567.032)}{\frac{22.260}{14}} = 142.73 > F_{0.05;1,14}(= 4.60)$

x_2 : F-to-remove $= \frac{SS_{Reg}(x_1,x_2)-SS_{Reg}(x_1)}{MS_{Res}(x_2,x_1)}$

x_2 : F-to-remove $= \frac{(793.975-775.482)}{\frac{22.260}{14}} = 11.63 > F_{0.05;1,14}(= 4.60)$

The smallest F-to-remove value is observed for x_2. Since it is significant, x_2 is retained in the model.

step 4: For the variables not yet in the model the following F-to-enter values are calculated:

x_3 : F-to-enter $= \frac{SS_{Reg}(x_1x_2,x_3)-SS_{Reg}(x_1x_2)}{MS_{Res}(x_3,x_1x_2)}$

x_3 : F-to-enter $= \frac{(795.834-793.975)}{\frac{20.401}{13}} = 1.18 < F_{0.05;1,13}(= 4.67)$

x_3 does not improve significantly the regression. Consequently it is not included. Since in a further step no variable can be added or deleted the procedure stops and the final regression equation is:
$\hat{y} = -42.001 + 0.777\,x_1 + 0.569\,x_2$.

In section 5.5 we will apply the SMLR to select the variables in different real-life datasets.

5.3.3   Principal component regression (PCR)

The Principal component regression (PCR) is a simple extension of the principal component analysis (PCA) and the multiple linear regression(MLR) [31, 34, 40, 54, 55]. In the first step, the principal components are calculated. The original variables are replaced by principal components (Pc), which are linear combinations of the columns in matrix $\mathbf{X}$. Most multivariate analysis textbooks (for example [4, 13, 68]) describe matrix methods for performing PCA. The goal is to find the eigenvectors of the covariance matrix. These eigenvectors correspond to the directions of the principal components of the original data. Their statistical significance is given by their corresponding eigenvalues. In more details, these techniques can be structured as [19]:

1. Subtract for each column, the mean of the column from each individual elements, resulting in a zero mean of the transformed variables and hence eliminating the need for a constant term in the regression model.

2. Calculate the covariance matrix $\mathbf{C}$.

$$\mathbf{C} = \frac{1}{n}\mathbf{X}^T\mathbf{X}.$$

3. Determine eigenvalues and eigenvectors of the matrix. $\mathbf{C}$ is real symmetric matrix so a positive real number λ and a nonzero vector α can be found such that:

$$\mathbf{C}\alpha = \lambda\alpha,$$

where λ is called an eigenvalue and α is an eigenvector of $\mathbf{C}$. To find a nonzero α the characteristic equation $\mid \mathbf{C} - \lambda\mathbf{I} \mid = 0$ must be solved. If $\mathbf{C}$ is a $n \times n$ matrix of full rank, n eigenvalues can be found $\lambda_1, \lambda_2, \ldots, \lambda_n$. Using $(\mathbf{C} - \lambda\mathbf{I}) = 0$ all the corresponding eigenvectors can be found.

4. Sort the eigenvalues (and corresponding eigenvectors) so that $\lambda_1 \geq \lambda_2 \geq \ldots \geq \lambda_n$.

5. Select the first $a \leq n$ eigenvectors and generate the dataset in the new representation.

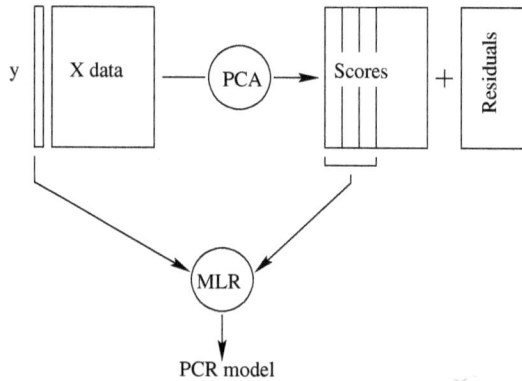

Figure 5.4: Principal component regression

The scores (values for ath principal component) of the most important principal components are then used as inputs for multiple linear regression (MLR) with the dependent variable $\mathbf{y}$ (see figure 5.4) [13, 27, 34]:

$$\mathbf{y} = \mathbf{Tb} + \mathbf{e}, \qquad (5.15)$$

where $\mathbf{T}$ is a $(m \times a)$ new matrix of data. By Analogy with MLR, the least squares solution for Eq. 5.15 is:

$$\mathbf{b} = (\mathbf{T}^T\mathbf{T})^{-1}\mathbf{T}^T\mathbf{y}. \qquad (5.16)$$

PCR solves the collinearity problems (by guaranteeing an invertible matrix $(\mathbf{T}^T\mathbf{T})$ in the calculation of $\mathbf{b}$ if $a << n$) and the ability to eliminate the less important principal components allows some noise reduction. The regression coefficients are more stable. This is due to the fact that the eigenvectors are orthogonal to each other. For PCR method the optimum number of principal components (a) corresponding to the PCR model has to be determined by cross-validation using an increasing number of components. The model with the smallest criterion error value on validation set can be regarded as the best model.

5.3.4 Partial least square regression (PLSR)

The PLSR method consists of a regression of the dependent variable y on the variables $t_1, t_2, \ldots$ which are latent variables (linear combinations of $x_1, x_2, \ldots, x_n$). However, in the PLSR method the latent variables are determined by using both y and the independent variables $x_1, x_2, \ldots, x_n$, whereas in the principal component regression (PCR) method, the latent variables (principal components) are determined using only the information coming from the independent variables. The PLSR method proceeds in an iterative manner and determines at each step a latent variable which is strongly connected to y, the force of the connection being measured by the importance of the covariance [13, 34, 40, 44, 54, 55, 80, 88].

The original and computationally simplest algorithm for the PLSR method was developed by Savante Wold, as given e.g. in Wold et al. [88]. It starts by finding the loading weight vector $\mathbf{w}_a$ for the ath latent variable by maximizing the covariance between the linear combination $\mathbf{X}_{a-1}\mathbf{w}_a$ and $\mathbf{y}_{a-1}$ under the constraint that $\mathbf{w}_a^T\mathbf{w}_a = 1$, where $\mathbf{X}_{a-1}$ and $\mathbf{y}_{a-1}$ are the old residuals and are calculated by subtracting the effects of the previous $(a-1)$th latent variables. This corresponds to finding the input vector $\mathbf{w}_a$ that maximizes the expression $\mathbf{w}_a^T\mathbf{X}_{a-1}^T\mathbf{y}_{a-1}$ i.e. the scaled covariance between $\mathbf{X}_{a-1}$ and $\mathbf{y}_{a-1}$. The steps of the PLSR algorithm are [54]:

1. Center the input variables $\mathbf{X}$ and $\mathbf{y}$ firstly. Choose A_{max} the number of latent variables and for each latent variable $a = 1, \ldots, A_{max}$ perform steps 2-6. Before step 2, we fix $\mathbf{X}_0 = \mathbf{X}$ and $\mathbf{y}_0 = \mathbf{y}$.

2. Starts by estimating the loading weight vector $\mathbf{w}_a$ for the ath latent variable, as the vector that maximizes the expression $\mathbf{w}_a^T\mathbf{X}_{a-1}^T\mathbf{y}_{a-1}$:

$$\mathbf{w}_a = \frac{\mathbf{X}_{a-1}^T\mathbf{y}_{a-1}}{\parallel \mathbf{X}_{a-1}^T\mathbf{y}_{a-1} \parallel}. \qquad (5.17)$$

3. Estimate the factor scores $\mathbf{t}_a$ as the projection of $\mathbf{X}_{a-1}$ on $\mathbf{w}_a$:

$$\mathbf{X}_{a-1} = \mathbf{t}_a \mathbf{w}_a + \mathbf{E}; \qquad (5.18)$$

the solution is (since $\mathbf{w}_a^T \mathbf{w}_a = 1$):

$$\mathbf{t}_a = \mathbf{X}_{a-1} \mathbf{w}_a. \qquad (5.19)$$

4. Regress $\mathbf{X}_{a-1}$ on $\mathbf{t}_a$ to find the loading vector $\mathbf{p}_a$:

$$\mathbf{X}_{a-1} = \mathbf{t}_a \mathbf{p}_a^T + \mathbf{E}, \qquad (5.20)$$

which gives the least square solution:

$$\mathbf{p}_a = \frac{\mathbf{X}_{a-1}^T \mathbf{t}_a}{\mathbf{t}_a^T \mathbf{t}_a}. \qquad (5.21)$$

5. In order to make estimations of $\mathbf{y}_{a-1}$ from $\mathbf{t}_a$ possible, the regression coefficient q_a for the ath latent variable is needed, which is determined by regression of $\mathbf{y}_{a-1}$ on $\mathbf{t}_a$:

$$\mathbf{y}_{a-1} = \mathbf{t}_a \, q_a + \mathbf{f}, \qquad (5.22)$$

which gives the solution:

$$q_a = \frac{\mathbf{y}_{a-1}^T \mathbf{t}_a}{\mathbf{t}_a^T \mathbf{t}_a}. \qquad (5.23)$$

6. New residuals $\mathbf{X}_a$ and $\mathbf{y}_a$ are calculated by subtracting the effect of the previous latent variables:

$$\mathbf{E} = \mathbf{X}_{a-1} - \mathbf{t}_a \mathbf{p}_a^T, \qquad (5.24)$$

$$\mathbf{f} = \mathbf{y}_{a-1} - \mathbf{t}_a \mathbf{q}_a. \qquad (5.25)$$

Replace the former $\mathbf{X}_{a-1}$ and $\mathbf{y}_{a-1}$ by the new residuals $\mathbf{E}$ and $\mathbf{f}$ and increase a by 1:

$$\mathbf{X}_a = \mathbf{E}, \qquad (5.26)$$

$$\mathbf{y}_a = \mathbf{f}, \qquad (5.27)$$

$$a = a + 1.$$

7. Determine A, the optimal number of latent variables to retain in the calibration model by cross-validation.

8. Similar to PCR, the regression coefficients $\mathbf{b}_{PLS}$ are useful for the interpretation of the PLSR model and for predictions of validation samples ($\mathbf{X}_{val}$) as $\mathbf{y} = \mathbf{X}_{val}\mathbf{b}_{PLS}$. The $\mathbf{b}_{PLS}$ coefficients are calculated after A latent variables as:

$$\mathbf{b}_{PLS} = \mathbf{W}(\mathbf{P}^T\mathbf{W})^{-1}\mathbf{q}, \tag{5.28}$$

where $\mathbf{W}$ is $(\mathbf{w}_1 \mid \mathbf{w}_2 \mid \ldots \mid \mathbf{w}_A)$, $\mathbf{P} = (\mathbf{p}_1 \mid \mathbf{p}_2 \mid \ldots \mid \mathbf{p}_A)$ and $\mathbf{q}^T = (q_1, \ldots, q_A)$.

This algorithm is also called the orthogonalized PLSR algorithm, since the estimated score and weight vectors are orthogonal, i.e, $\mathbf{t}_i^T\mathbf{t}_j = 0$ and $\mathbf{w}_i^T\mathbf{w}_j = 0$ where $i \neq j$.

5.4 Spectrophotometric variable selection by non-linear models

All these methods (SMLR, PCR and PLSR) make the assumption of the existence of a linear relation between the selected built variables on one hand, and the characteristic to be predicted on the other hand. This can obviously not be the case in the reality of certain applications, leading to the need for using non-linear models instead of linear ones.

The weakness of the stepwise multiple linear regression (SMLR) method is that it only describes the ability to model the training data, rather than the ability to predict with new samples [84].

In the literature, some authors use the PCA or PLS score vectors as inputs to non-linear models [27]. In the case of compressing data with PCA, one must be aware of some theoretical limitation. PCA is a linear projection method that fails to preserve the structure of a non-linear dataset. If there is some non-linearity in $\mathbf{X}$ (or between $\mathbf{X}$ and $\mathbf{y}$), this non -linearity can appear as a small perturbation on a linear solution and will not be described by the first principal components as in linear case. Concerning the latent variables, they are designed to conserve information linearly correlated with the response (dependent variable) and some non-linear relevant information might be rejected in higher order latent variables that are not retained in the model [22]. For this reason, it is not recommended to pre-process data with PCA or PLS before non-linear models for example artificial neural networks.

It should be noted that other types of models may be used to analyze high-dimensional data: the Functional Data Analysis (FDA) methods. In FDA, data are considered as infinite-dimensional vectors (or functions). This may be particularly advantageous when the data consists in samplings of a continuous physical phenomenon. FDA has recently be applied to spectroscopic data [29]. These recent methods will not be compared to our approach in this book because this would require another complete study. A general and complete introduction to *Functional Data Analysis* can be found in Ramsay et al. [48].

5.4.1 Forward-backward selection by RBF networks

Given these limitations, we propose a method for variable selection based on the three following principles:

1. Use of a *non-linear regression model* (artificial neural network);

2. Choice of the variables based on an *incremental procedure* (forward-backward selection);

3. Choice of the variables according to an error criterion computed on *a validation dataset.*

The combination of these three principles will lead respectively to enhanced calibration capabilities (compared to linear methods), to an efficient compromise between inefficient and exhaustive searches of the variables to select, and to an objective assessment of the performances (and objective comparisons between methods as a corollary). It should be noted that the non-linear model used in the method of variable selection is a RBF network with the linear and constant terms (Eq. 2.2). The linear part of the Eq. 2.2 can handle the linear part and the classical

RBF network builds the non-linear part of the model. The use of a RBF network model with linear and constant terms can be interesting with infrared spectrophotometric data where the non-linear effects observed generally correspond to deviation from linear solution [22].

We propose a method of spectral data selection based on a criterion of validation known as *forward-backward selection* (FBS). The selection of the spectral data is divided into two stages [10, 9]:

1. The first stage is the forward selection. It starts with the construction of the n possible models each using one single variable. We calculate the error criterion for each of these models and we choose the one that minimizes the criterion. This leads to the choice of the first variable. Secondly, we keep this variable, and build $n - 1$ models by adding one of the remaining spectral variables. The error criterion for each one of these models is calculated and we choose the model that minimizes this criterion. A second variable is then selected. We continue this process until the value of the error criterion increases. As detailed below, it is therefore necessary to evaluate the error criterion on a *validation dataset*, independent from the training dataset. By validation dataset, we mean a set of samples (observations) not used for training (fitting the calibration model). Depending on the research discipline, some authors use the words 'external validation set', 'external set' or 'prediction set'; the important concept is that the samples used to validate a method must be independent from those used for training, regardless of the terminology. Only the use of a validation dataset will ensure an objective evaluation of the error resulting from each model. Moreover, only the error on a validation dataset will increase when the number of selected variables is too large, leading to the well-known overfitting phenomenon (see figure 5.5) [41].

2. The second stage is the backward selection. It consists of eliminating the least significant spectral data already selected in the first stage. If u spectral variables were selected after the first stage, u models are built by removing one of the selected variables. The error criterion is calculated on each of these models, and the one

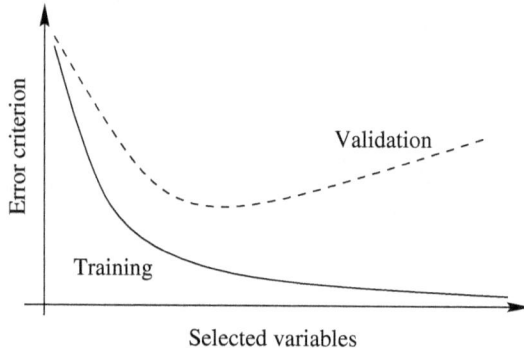

Figure 5.5: typical evolution of the performances of training and validation.

with the lower error is selected. Once the model is chosen, we compare its error to the error of the model obtained at the preceding stage. If the new error criterion is lower, then the eliminated spectral variable is not significant and may be removed. The process is then repeated on the remaining spectral variables. The backward selection is stopped when the lower error among all models calculated at one step is higher than the error at the previous step.

It should be noted that we can also use a linear model instead of non-linear RBF networks in the incremental procedure of variable selection. In section 5.5, we will compare the use of a linear and of a non-linear model. In the case of a linear model we speak about FBS-lin, whereas in the case of a non-linear model we speak about FBS-RBFN.

5.4.2 Error criterion

As mentioned above, at each step of the forward-backward selection algorithm, the error of several models must be evaluated on data independent from the ones used for learning. This is achieved through the use of a validation set V containing N_V spectra:

$$V = \{(\mathbf{x}_q, y_q) \in R^d \times R,\ 1 \leq q \leq N_V : y_q = f(\mathbf{x}_q)\}. \qquad (5.29)$$

The error criterion can be chosen as the normalized mean square error defined as [86]:

$$NMSE_V = \frac{\frac{1}{N_V}\sum_{q=1}^{N_V}(\hat{y}_q - y_q)^2}{\frac{1}{N_T+N_V}\sum_{j=1}^{N_T+N_V}(y_j - \bar{y})^2}, \qquad (5.30)$$

where N_T, N_V are the number of samples included in the training set and the validation set respectively, $\hat{y}_q$ is the value predicted by the model and y_q is the actual value corresponding to spectrum q. Note that equation 5.30 normalizes the errors with respect to the standard deviation of y values in the combined learning and validation sets, the reason being to use as much data as possible to estimate this standard deviation. As this estimation does not depend on the model, the comparison of performances between models remains objective, whatever is the set used to estimate this standard deviation.

5.5 Real-life examples

In this section, we use our procedure of variable selection with a linear model (FBS-lin) and a non-linear model (FBS-RBFN) proposed in section 5.4, and the SMLR, PCR and PLSR methods detailed in section 5.3, to retain only the most relevant variables in four real-life datasets.

The first three datasets (wine dataset, orange juice dataset and milk powder dataset) as well as their distribution between training and validation sets were provided by the laboratory of spectrophotometry of the research unit AGRO/BNUT of UCL, whereas the fourth dataset (apple dataset) as well as its distribution (between training and validation sets) was provided by ENITIAA/INRA (unit of Sensometry and Chemometrics) in Nantes, France.

5.5.1 Wine dataset

The first dataset relates to the determination of alcohol concentration by mid-infrared spectroscopy in wine samples. The training and validation sets contain 94 and 30 spectra respectively, with 256 spectral variables that are the absorbance ($\log \frac{1}{T}$) at 256 wavenumbers between 4000 and 400 cm^{-1} (where T is the light transmittance through the sample thickness). Figure 5.6(a) shows a collection of 94 wine spectra used in the training and figure 5.6(b) shows a collection of 30 wine spectra used in the prediction.

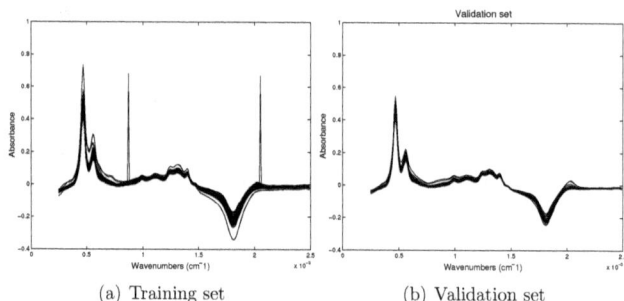

| (a) Training set | (b) Validation set |

Figure 5.6: Mid-infrared transmittance spectra of wine.

In table 5.4 the predictive ability of each of the PCR, PLSR, SMLR, FBS-lin and FBS-RBFN models is compared in terms of normalized mean square error on a validation set (Eq. 5.30). The first three methods were detailed in section 5.3, while the last two ones correspond to the forward-backward selection procedure detailed in section 5.4, using respectively a linear model and a non-linear one (RBF network) for the prediction. Table 5.4 shows the error criterion $NMSE_V$ on the validation set for each procedure, together with the number of variables or principal components or latent variables involved in each model.

In the PCR model the lowest $NMSE_V$ value was obtained with

Calibration model	Number of variables	$NMSE_V$
PCR	20	0.0105
PLSR	11	0.0116
SMLR	14	0.0039
FBS-lin	17	0.0012
FBS-RBFN	**20**	**0.0009**

Table 5.4: Results of prediction on the wine dataset for the five procedures.

20 principal components. Similarly for the PLSR model the lowest $NMSE_V$ value was obtained with 11 latent variables. In the FBS-lin case, the initial number of selected variables using the forward selection is 20, but after the application of the backward selection this number was reduced to 17, as illustrated in figure 5.7(a). It can be seen in this figure that the backward procedure could have been stopped at 19 variables, leading to a smaller error criterion ($NMSE_V = 0.001129$). We chose here to continue the procedure until the $NMSE_V$ reaches the same value as at the end of the forward stage, thus privileging a lower number of variables. Regarding the FBS-RBFN procedure, we tested Radial-Basis Function Networks with 2 to 8 centers (neurons) in the hidden layer. On the wine database, the best result was obtained with 3 centers in the hidden layer; the width scaling factor (WSF) was also optimized on the validation set ($WSF_{opt} = 1.03$). Also in this case, the initial number of variables selected by the forward selection is 32, but after the backward selection this number was reduced to 20, as shown in figure 5.7(b). In Figures 5.7(a) and 5.7(b) the validation $NMSE_V$ is shown as a function of the number of forward-backward selected variables. Note that the set of variables obtained by FBS-lin is different from the set of variables obtained by FBS-RBFN.

The variables selected by the FBS-lin method are:

$$subset = \begin{bmatrix} \mathbf{x}_{71} \; \mathbf{x}_{26} \; \mathbf{x}_{16} \; \mathbf{x}_{63} \; \mathbf{x}_4 \; \mathbf{x}_{59} \; \mathbf{x}_{58} \; \mathbf{x}_{85} \; \mathbf{x}_{14} \; \mathbf{x}_{81} \; \mathbf{x}_{70} \; \mathbf{x}_{182} \; \mathbf{x}_{99} \; \mathbf{x}_{172} \; \mathbf{x}_{90} \; \mathbf{x}_{11} \\ \mathbf{x}_{206} \end{bmatrix},$$

and the variables selected by the FBS-RBFN method are:

$$subset = \begin{bmatrix} \mathbf{x}_{177} \; \mathbf{x}_{26} \; \mathbf{x}_{16} \; \mathbf{x}_{134} \; \mathbf{x}_{129} \; \mathbf{x}_{153} \; \mathbf{x}_{65} \; \mathbf{x}_{29} \; \mathbf{x}_{115} \; \mathbf{x}_{60} \; \mathbf{x}_{64} \; \mathbf{x}_{191} \; \mathbf{x}_{200} \; \mathbf{x}_{152} \\ \mathbf{x}_{198} \; \mathbf{x}_{50} \; \mathbf{x}_{76} \; \mathbf{x}_3 \; \mathbf{x}_{10} \; \mathbf{x}_1 \end{bmatrix}.$$

Errors obtained with the forward-backward selection (both linear and non-linear calibration models) are much lower than the errors obtained with the other procedures. The use of RBFN only slightly improves the results compared to the use of a linear model, because the relationship between the absorbance $\log \frac{1}{T}$ and the concentration is approximately linear according to the Lambert-Berr's law in mid-infrared spectroscopy [3]. Figures 5.8(a), 5.8(b), 5.8(c), 5.8(d) and 5.8(e) show the relationship between the predicted alcohol concentration and the actual concentration with the SMLR, PCR, PLSR, FBS-lin and FBS-RBFN methods of variable selection.

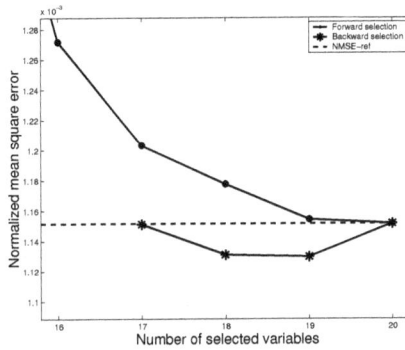

(a) FBS-lin method

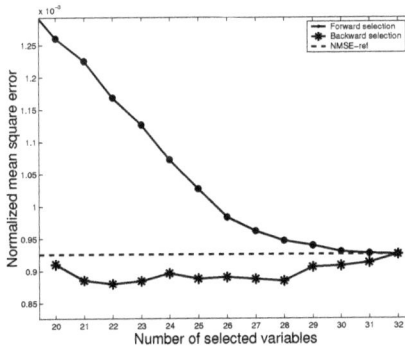

(b) FBS-RBFN method

Figure 5.7: $NMSE_V$ with respect to the number of selected variables.

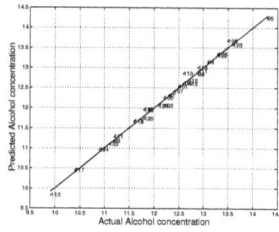

(a) SMLR method

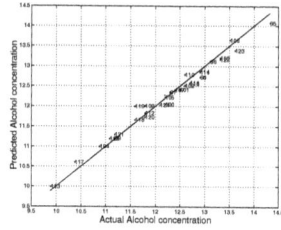

(b) PCR method

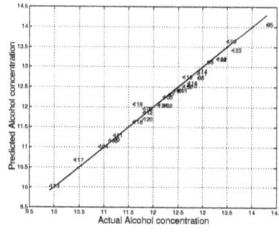

(c) PLSR method

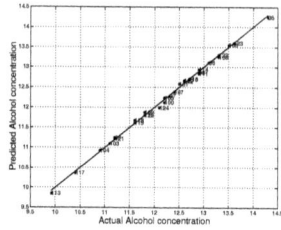

(d) FBS-lin method

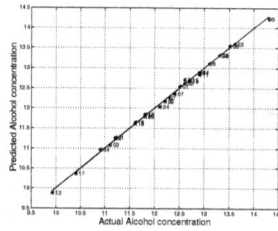

(e) FBS-RBFN method

Figure 5.8: Predicted alcohol concentration with respect to the measured alcohol concentration in wine.

5.5.2   Orange juice dataset

The second data set relates to the determination of sugar (saccharose) by near-infrared reflectance spectroscopy in orange juice samples. In this case, training and validation sets contain respectively 150 and 68 spectra, with 700 spectral variables that are the absorbance ($\log \frac{1}{R}$) at 700 wavelengths between 1100 and 2500 nm (where R is the light reflectance on the sample surface). Figure 5.9(a) shows a collection of 150 orange juice spectra used in the training set and figure 5.9(b) shows a collection of 68 orange juice spectra used in the prediction set.

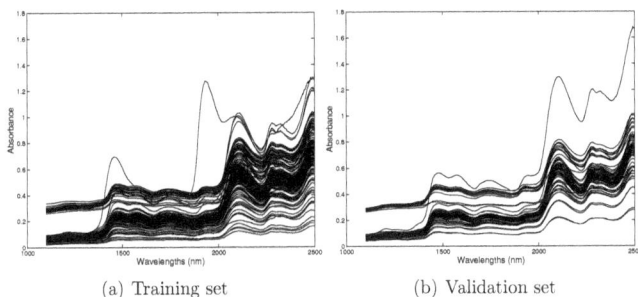

(a) Training set   (b) Validation set

Figure 5.9: Near-infrared reflectance spectra of orange juice.

On this second dataset (orange juice samples), we used the five methods already used for the wine dataset, and added two methods: PCA-RBFN and PLS-RBF, where the variable selection takes place exactly as in PCR and PLSR respectively (thus using a linear model for the selection). Nevertheless, once the variables are selected, a non-linear model (RBFN) is used to predict the sugar concentration (saccharose). To determine the optimal number of principal components and latent variables in the PCA-RBFN and PLS-RBFN methods, we tested Radial Basis Function Networks with various values of principal components and latent variables. The number of principal components and of latent variables used varies respectively from 4 to 35 and from 1 to 25. In the FBS-RBFN procedure, we tested Radial Basis Function Networks with different number of centers (2 to 11) in the hidden layer.

Table 5.5 gives a comparison between the predictive ability of the
PCR, PLSR, SMLR, FBS-lin, PCA-RBFN, PLS-RBFN and FBS-RBFN
models in terms of normalized mean square error $(NMSE_V)$ on the
validation set.

Calibration model	Number of variables	$NMSE_V$
PCR	42	0.2443
PLSR	16	0.2360
SMLR	16	0.5137
FBS-lin	7	0.2265
PCA-RBFN	12	0.1407
PLS-RBFN	14	0.1364
FBS-RBFN	**13**	**0.0691**

Table 5.5: Results of prediction on the orange juice dataset for the seven procedures.

Figures 5.10(a) and 5.10(b) show the evolution of the normalized
mean square error $(NMSE_V)$ according to the number of principal
components and of latent variables, on the training and validation
sets.   In the case of the PCA-RBFN method, the optimal number
of principal components is equal to 12 with a RBFN model made
up of 5 centers in the hidden layer. In the case of the PLS-RBFN
method, the optimal number of latent variables is equal to 14 with
a model RBFN made up of 26 centers in the hidden layer. Figure
5.11 shows the evolution of the error criterion $(NMSE_V)$ according
to the number of centers in the case of the FBS-RBFN method. The
smallest error is obtained using our procedure, the FBS-RBFN calibration model, with 8 centers in the hidden layer and, a $WSF_{opt} = 5.95$.

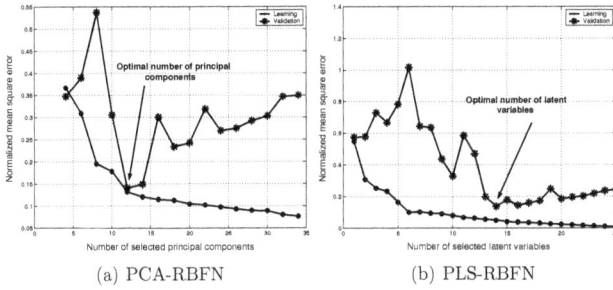

(a) PCA-RBFN   (b) PLS-RBFN

Figure 5.10: $NMSE_V$ with respect to the number of principal components and of latent variables for the orange juice dataset.

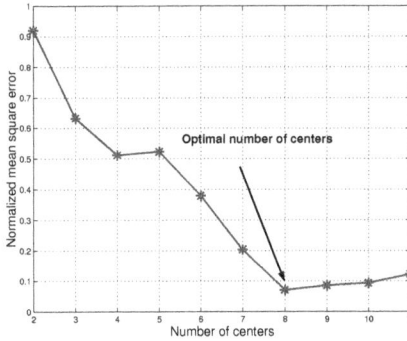

Figure 5.11: $NMSE_V$ with respect to the number of centers for the orange juice dataset.

Table 5.5 shows clearly that the calibration models using non-linear methods (RBFN) largely reduce the errors.

Figure 5.12(a) shows that in the FBS-lin case the backward selec-

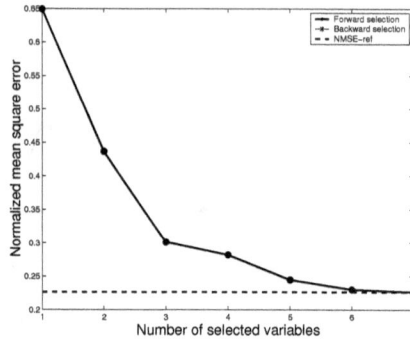

(a) FBS-lin method

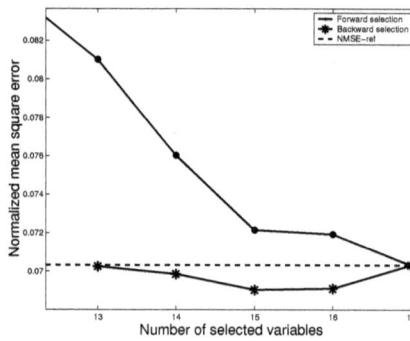

(b) FBS-RBFN method

Figure 5.12: $NMSE_V$ with respect to the number of selected variables for the orange juice samples.

tion did not reduce the number of variables previously selected by the forward stage (7 variables). On the contrary, figure 5.12(b) shows the effect of the backward selection on the reduction of the number of variables previously selected by the forward stage in the case of the

FBS-RBFN. The number of variables is reduced from 17 to 13 for the same level of $NMSE_V$.

Note also in this example that the set of variables obtained by FBS-lin is different from the set of variables obtained by FBS-RBFN.

The variables selected by the FBS-lin method are:

$subset = [\mathbf{x}_{494}\, \mathbf{x}_{521}\, \mathbf{x}_{590}\, \mathbf{x}_{681}\, \mathbf{x}_{588}\, \mathbf{x}_{595}\, \mathbf{x}_{682}]$,

and the variables selected by the FBS-RBFN method are:

$subset = [\mathbf{x}_{256}\, \mathbf{x}_{383}\, \mathbf{x}_{282}\, \mathbf{x}_{346}\, \mathbf{x}_{384}\, \mathbf{x}_{305}\, \mathbf{x}_{319}\, \mathbf{x}_{286}\, \mathbf{x}_{254}\, \mathbf{x}_{347}\, \mathbf{x}_{277}\, \mathbf{x}_{332}\, \mathbf{x}_{337}]$.

The predicted saccharose concentration according to the actual saccharose concentration with PCR, PLSR, FBS-lin, PCA-RBFN, PLS-RBFN and FBS-RBFN calibration model is presented respectively in figures 5.13(a), 5.13(b), 5.13(c), 5.14(a), 5.14(b) and 5.14(c). These figures show the improvement obtained with the use of forward-backward selection procedure by non-linear models (RBF networks) compared to other methods, as measured in table 5.5.

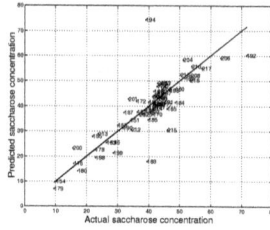

(a) PCR method

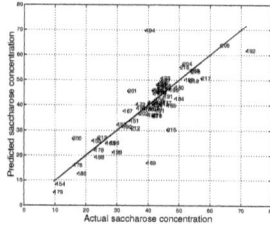

(b) PLSR method

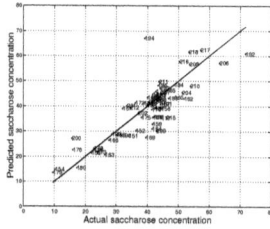

(c) FBS-lin method

Figure 5.13: Predicted saccharose concentration with respect to measured saccharose concentration with linear methods on the orange juice samples.

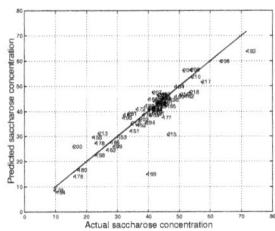

(a) PCA-RBFN method

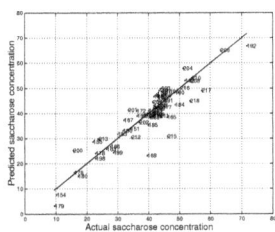

(b) PLS-RBFN method

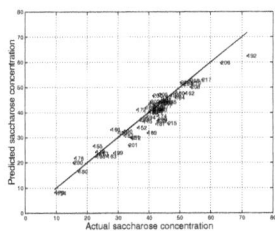

(c) FBS-RBFN method

Figure 5.14: Predicted saccharose concentration with respect to measured saccharose concentration with non-linear methods on the orange juice samples.

5.5.3   Milk powder dataset

The third dataset consists of near-infrared reflectance spectra of milk powder obtained in the wavelength range from 1100 to 2500 nm at regular intervals of 2 nm. The y value to predict is the water content in a milk powder. In this example the number of samples is too small compared to the number of spectral data (variables); therefore it is necessary to select the most relevant variables in order to obtain a simple model. The training and validation sets consist respectively of 27 and 10 samples, with 700 spectral variables that are the absorbance ($\log \frac{1}{R}$) at 700 wavelengths. Figure 5.15 shows the spectra of milk powder used in the training and validation sets.

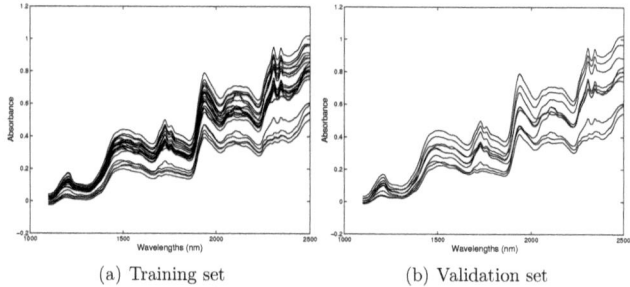

(a) Training set   (b) Validation set

Figure 5.15: Near-infrared reflectance spectra of milk powder.

On this third dataset (samples of the milk powder) the number of samples is much smaller than the number of variables; therefore we used the seven methods already used for the orange juice dataset to select the relevant variables. Table 5.6 shows the $NMSE_V$ errors obtained as well as the number of selected variables by the seven variable selection methods.

In the PCR method the smallest $NMSE_V$ value was obtained

Calibration model	Number of variables	$NMSE_V$
PCR	10	0.9250
PLSR	7	0.8758
SMLR	4	1.5288
FBS-lin	4	0.7900
PCA-RBFN	5	0.6638
PLS-RBFN	3	0.6478
FBS-RBFN	**6**	**0.5309**

Table 5.6: Results of prediction on the milk powder dataset for the seven procedures.

with 10 principal components. Similarly for the PLSR model, the lowest $NMSE_V$ value was obtained with 7 latent variables. Figure 5.16(a) shows that the number of selected variables by the FBS-lin method is 4, while the backward selection did not reduce the number of variables previously selected by the forward selection. In the case of the PCA-RBFN method, the optimal number of principal components is equal to 5 with a RBFN model made up of 6 centers in the hidden layer. However in the case of the PLS-RBFN method, the optimal number of latent variables is equal to 3 with a RBFN model made up of 5 centers in the hidden layer. Regarding the FBS-RBFN method, we tested Radial Basis Function Networks with 2 to 8 centers (neurons) in the hidden layer. In this dataset, the smallest $NMSE_V$ was obtained with 3 centers in the hidden layer and with a $WSF_{opt} = 5.95$, as illustrated in figure 5.17. The number of selected variables by the FBS-RBFN method is 6 (see figure 5.16(b)).

We also notice that the variable selection methods using non-linear models give better results and that the best model obtained with our incremental procedure is the FBS-RBFN.

Figures 5.18(a), 5.18(b), 5.18(c), 5.19(a), 5.19(b) and 5.19(c) show

the relationship between the predicted and actual water content in milk powder with PCR, PLSR, FBS-lin, PCA-RBFN, PLS-RBFN and FBS-RBFN variable selection calibration models.
In this dataset, the variables selected by the FBS-lin method are:

$$subset = \begin{bmatrix} \mathbf{x}_{417} \ \mathbf{x}_{420} \ \mathbf{x}_{415} \ \mathbf{x}_{411} \end{bmatrix},$$

and the variables selected by the FBS-RBFN method are:

$$subset = \begin{bmatrix} \mathbf{x}_{572} \ \mathbf{x}_{382} \ \mathbf{x}_{505} \ \mathbf{x}_{57} \ \mathbf{x}_{56} \ \mathbf{x}_{263} \end{bmatrix}.$$

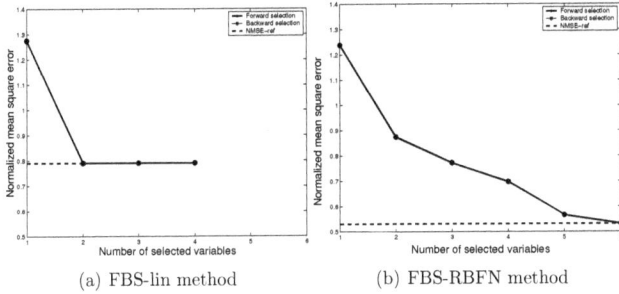

(a) FBS-lin method   (b) FBS-RBFN method

Figure 5.16: $NMSE_V$ with respect to the number of selected variables for the milk powder dataset.

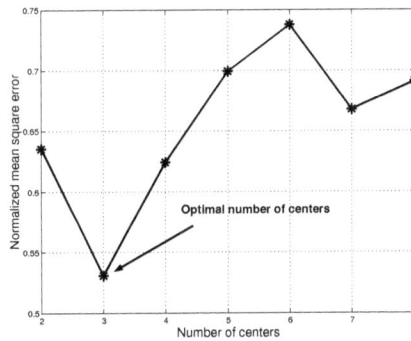

Figure 5.17: $NMSE_V$ with respect to the number of centers for the milk powder dataset.

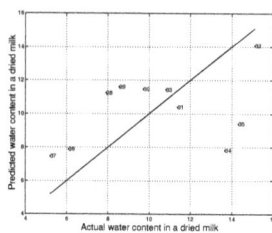

(a) PCR method

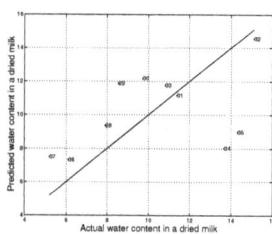

(b) PLSR method

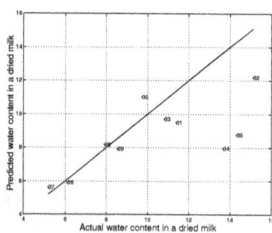

(c) FBS-lin method

Figure 5.18: Predicted water with respect to measured water with linear methods in milk powder samples.

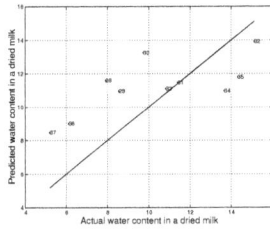

(a) PCA-RBFN method

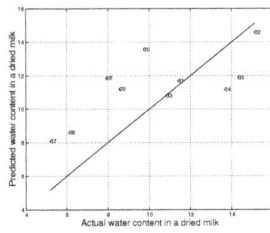

(b) PLS-RBFN method

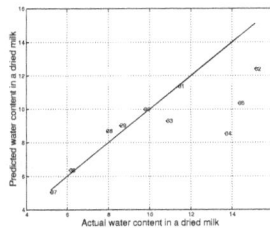

(c) FBS-RBFN method

Figure 5.19: Predicted water with respect to measured water with non-linear methods in milk powder samples.

5.5.4 Apple dataset

The fourth dataset concerns the near-infrared reflectance spectra of apple, and the prediction of the crushing force that should be employed to insert an object (stem or marble) in the skin of an apple to determine if it is too ripe or not. This force is proportional to the firmness of the flesh. If an apple is not too ripe and is well crunching, it will be firm and the crushing force will be high. The apple spectra are measured by near-infrared reflectance spectroscopy. The training and validation sets contain respectively 225 and 112 spectra, with 110 spectral variables that are the absorbance ($\log \frac{1}{R}$) at 110 wavelengths between 1100 and 2190 nm (where R is the light reflectance on the sample surface). The spectra used in the training and the prediction are illustrated by figure 5.20.

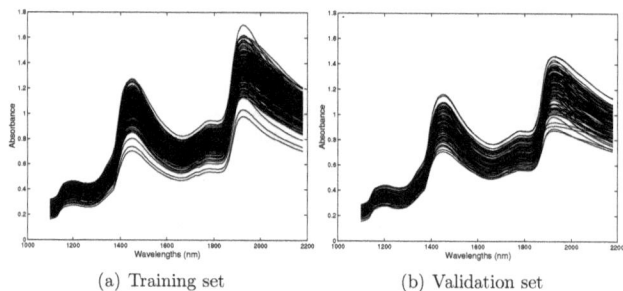

(a) Training set   (b) Validation set

Figure 5.20: Near-infrared reflectance spectra of apples.

Since the number of samples is larger than the number of variables in this dataset, we used the multiple linear regression (MLR) and Radial Basis Functions Networks (RBFN) to predict the crushing force with all variables. We also used the seven variable selection methods already used in the previous dataset to select the relevant variables. Table 5.7 shows the $NMSE_V$ errors obtained by multiple linear regression as well as the number of selected variables by the seven variable selection methods.

Calibration model	Number of variables	$NMSE_V$
MLR	110	0.9830
RBFN	110	0.6258
PCR	7	0.6721
PLSR	4	0.6029
SMLR	7	0.4417
FBS-lin	2	0.7211
PCA-RBFN	14	0.4143
PLS-RBFN	20	0.3856
FBS-RBFN	**8**	**0.2787**

Table 5.7: Results of prediction on the apple dataset for the nine procedures.

The smallest $NMSE_V$ value was obtained with 7 principal components in the case of the PCR calibration model. Similarly, for the PLSR calibration model, the smallest $NMSE_V$ value was obtained with 4 latent variables. In the FBS-lin case, the best model is obtained only with two variables, as illustrated in figure 5.21(a). In the case of the PCA-RBFN model, the lower $NMSE_V$ value is obtained for six centers in the hidden layer with 14 principal components as inputs, whereas in the case of PLS-RBFN, the lower $NMSE_V$ value is obtained for eight centers in the hidden layer with 20 latent variables as inputs. Regarding the FBS-RBFN calibration model, we tested Radial Basis Functions Networks with 2 to 10 centers (neurons) in the hidden layer (see figure 5.22). The smallest $NMSE_V$ value is obtained for a RBF network having eight selected variables as inputs, six centers in the hidden layer and a $WSF_{opt} = 1.4$, as illustrated in figure 5.21(b).

According to table 5.7 we can conclude that the variable selection methods using non-linear models give the best results, and that the best model is obtained with our incremental variable selection procedure the FBS-RBFN. Figures 5.23(a), 5.23(b), 5.23(c), 5.24(a), 5.24(b) and

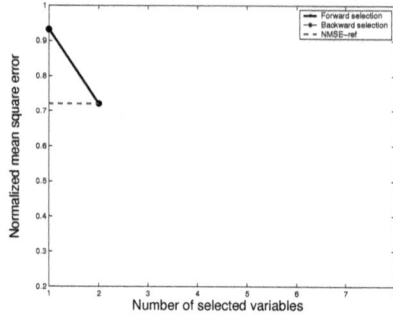

(a) FBS-lin method

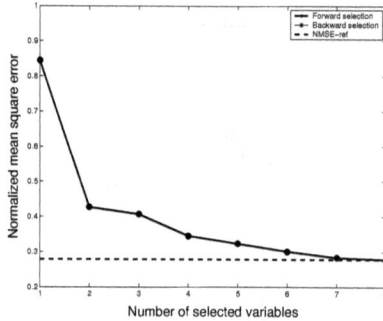

(b) FBS-RBFN method

Figure 5.21: $NMSE_V$ with respect to the number of selected variables for the apple dataset.

5.24(c) show the relationship between predicted and actual crushing force for the apple samples with PCR, PLSR, FBS-lin, PCA-RBFN, PLS-RBFN and FBS-RBFN variable selection calibration models.

In this dataset the variables selected by the FBS-lin method are:

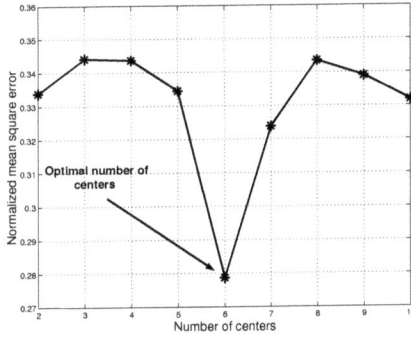

Figure 5.22: $NMSE_V$ with respect to the number of centers for the apple dataset.

$$subset = \begin{bmatrix} \mathbf{x}_{84} & \mathbf{x}_{85} \end{bmatrix},$$

and the variables selected by the FBS-RBFN method are:

$$subset = \begin{bmatrix} \mathbf{x}_{84} & \mathbf{x}_{37} & \mathbf{x}_{79} & \mathbf{x}_{106} & \mathbf{x}_{85} & \mathbf{x}_{83} & \mathbf{x}_{38} & \mathbf{x}_{93} \end{bmatrix}.$$

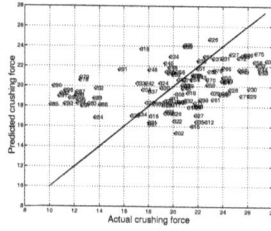

(a) PCR method

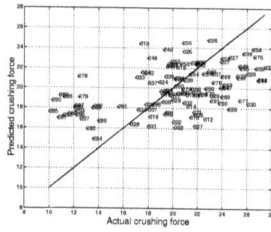

(b) PLSR method

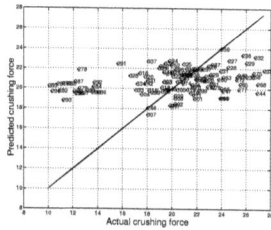

(c) FBS-lin method

Figure 5.23: Predicted crushing force with respect to the measured crushing force with linear methods in the apple samples.

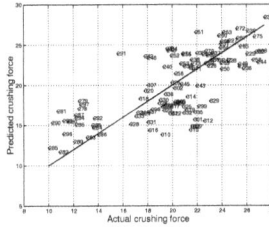

(a) PCA-RBFN method

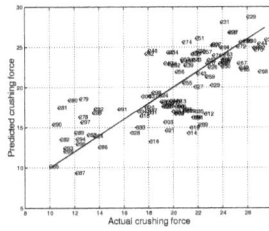

(b) PLS-RBFN method

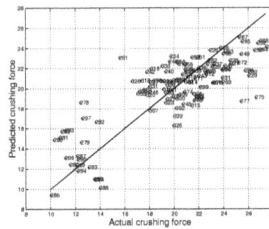

(c) FBS-RBFN method

Figure 5.24: Predicted crushing force with respect to the measured crushing force with non-linear methods in the apple samples.

5.6 Improvements

In this section four improvements of the previous method are presented. Firstly, we use the cross-validation method to select the relevant variables with the FBS-lin and PCR methods. Secondly, we use graphical information to see the distribution of the data and to detect the presence of outliers in the wine and orange juice datasets. Thirdly, we propose a sequential method to distribute data homogeneously between the training and validation sets. Finally, we suggest to improve the spectrophotometric variable selection method by combining it with the measure of mutual information between the dependent variable and independent variables.

5.6.1 Cross-validation

In the four real-life datasets, we used the external validation (hold-out method) to select the relevant variables. Since the number of samples in each dataset is not very large, it would have been more efficient to use a cross-validation method (random subsampling method) to estimate the generalization error, which is the average of the errors obtained on each randomly drawn validation set. To illustrate this, we used the variable selection procedure with the linear model FBS-lin and the PCR method on the wine dataset.

We gathered the 124 wine samples in only one set. Thereafter we drew randomly two-third (2/3) from the data to form the training set. The remaining third (1/3) data form the validation set. In the first experiment, the FBS-lin incremental procedure of variable selection is applied to each draw to select the relevant variables. The number of draws is fixed to 10. Table 5.8 shows the number of selected variables, and the normalized mean square error on validation set ($NMSE_V$) obtained for each draw in the case of the FBS-lin method. In the second experiment, we used the PCR method to select the principal components which will replace the original variables. It should be noted that the training and validation sets are the same as those used in the first experiment. Table 5.9 shows the optimal number of selected principal components, and the error on validation set obtained for each draw in the case of the PCR method.

Procedure of variable selection (FBS-lin)		
Draw	Number of variables	$NMSE_V$
1	24	0.00065
2	17	0.00067
3	19	0.0011
4	17	0.00086
5	23	0.00087
6	16	0.0010
7	23	0.00066
8	28	0.00042
9	15	0.00043
10	24	0.00061

Table 5.8: Number of selected variables and their corresponding $NMSE_V$ according to the draw for the wine dataset.

PCR method		
Draw	Number of principal components	$NMSE_V$
1	30	0.0023
2	18	0.0026
3	16	0.0086
4	27	0.0026
5	28	0.0016
6	30	0.0028
7	22	0.0020
8	16	0.0018
9	16	0.0029
10	14	0.0012

Table 5.9: Number of selected principal components and their corresponding $NMSE_V$ according to the draw for the wine dataset.

Several remarks can be made by analyzing these results:

- The number of selected variables and the number of principal components are different for each draw and strongly depend on the distribution of the data between the training and validation sets.

- We cannot calculate the generalization error, which is the average of the errors obtained in each draw, since the number of selected variables and principal components by each model are not identical.

- Even in the cases where the number of selected variables or the number of selected principal components is identical, the variables selected or the principal components are not identical. For example in table 5.8 the number of selected variables is the same one in draws 2 and 4, as was shown by these two subsets:

$$subset(2) = [x_{169} \, x_{31} \, x_{27} \, x_{70} \, x_{111} \, x_{78} \, x_{102} \, x_{23} \, x_{30} \, x_{35} \, x_{37} \, x_{128} \, x_{71} \\ x_{179} \, x_{12} \, x_{131} \, x_{178}],$$

$$subset(4) = [x_{189} \, x_{256} \, x_{27} \, x_{41} \, x_{15} \, x_{183} \, x_{115} \, x_{127} \, x_6 \, x_{43} \, x_{54} \, x_{250} \, x_1 \\ x_{188} \, x_{224} \, x_{111} \, x_{103}],$$

It should be mentioned that in the available literature, in the case of the PCR or the PLSR methods, several authors calculate the generalization error which is the average of the errors obtained in each draw, by supposing that the principal components or the latent variables are the same ones for each draw [12, 28, 30]. Faced with this difficulty, to avoid the necessity of using a cross-validation method, we propose in subsection 5.6.3 a method which makes it possible to distribute the data between training and validation sets.

5.6.2   Graphical detection of outliers

The visualization of data has always been very important in chemometrics, and it is impossible to discuss chemometrics without showing plots [28, 32]. One possibility is to simply plot all spectra on the same graph. Evident outliers will become apparent. It is also possible to identify noisy wavelength regions, and perhaps exclude them from the model [22].

Graphical display methods, mapping the objects (samples) from a high-dimensional to a two-dimensional space, are especially important in the early stage of data analysis. These methods often provide useful information about the relationships between samples in a dataset.

Principal component analysis (PCA) gives a linear reduction and is widely used for this purpose [22, 56, 76, 87].

Since PCA produces new variables (principal components), such that the highest amount of variance is explained by the first principal components (eigenvectors), the score plots can be used to give a good representation of the data. By using a small number of score plots (e.g. $Pc1 - Pc2$, $Pc1 - Pc3$ and $Pc2 - Pc3$), useful visual information can be obtained about the data distribution and the presence of outliers [22].

In what follows, we use this graphic representation to see the distribution of the data, and to detect the presence of outliers in the wine and orange juice datasets.

1. Wine dataset

All spectra of the wine dataset are plotted in figure 5.25. Spectra 34, 35 and 84, shown in figure 5.25 can be regarded as outliers. Figure 5.26 gives a typical example of a score plot for two first principal components ($Pc1 - Pc2$), after the application of the PCA on the 124 wine spectra. In figure 5.26, one dense region and few outliers can be seen, so that we can consider that samples 34, 35 and 84 are outliers and consequently can be eliminated from the wine dataset.

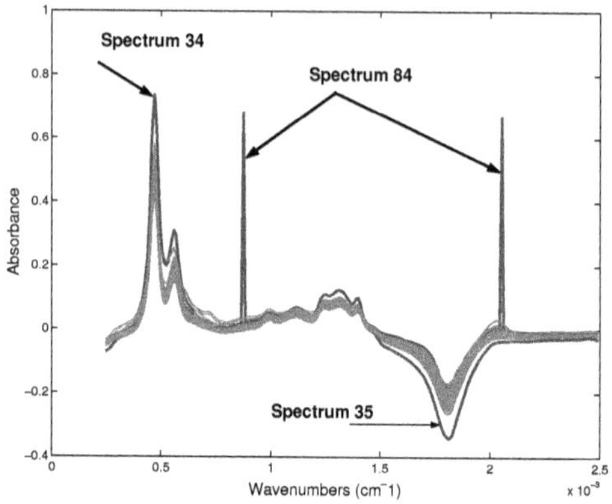

Figure 5.25: All the spectra of the wine dataset.

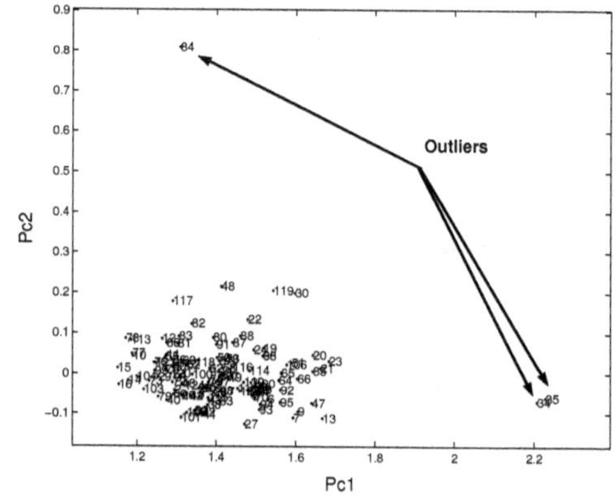

Figure 5.26: Score map (Pc1-Pc2) of the wine dataset.

2. Orange juice dataset

Figure 5.27 shows all the spectra of the orange juice dataset. We can regard spectra 130 and 194 illustrated in figure 5.27 as outliers. Figure 5.28 gives a typical example of a score plot for two first principal components $(Pc1 - Pc2)$, after the application of the PCA on the 218 orange juice spectra. In figure 5.28, two dense regions and few outliers can be seen, and we can consider that samples 130 and 194 are outliers, and can consequently be eliminated from the orange juice dataset.

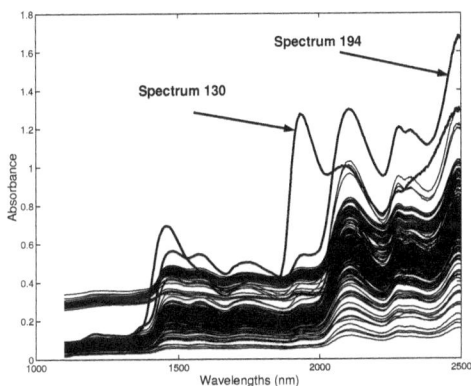

Figure 5.27: All the spectra of the orange juice dataset.

There is not doubt that PCA continues to play an important role, and is still a basic method in the display of multivariate dataset. In many cases, the inter-point distances in the display space reflect the trends in the original space. However, its drawback is also obvious. PCA imposes a linear structure on the variables that may obscure useful information contained in non-linear combinations. The first few principal components are related to the largest variance in the dataset. If an important variable contains a small amount of variance, this may not be

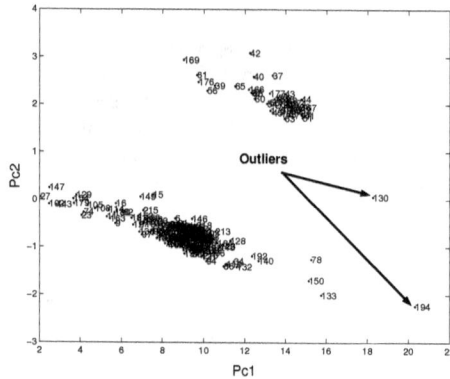

Figure 5.28: Score map (Pc1-Pc2) of the orange juice dataset.

reflected in the first few principal components.

Outliers in the **X**-space can be due to measurement or handling errors, in which case they should be eliminated. They can also be due to the presence of samples that belong to another population, to impurities in the outliers that are not present in the other samples, or extreme amounts of constituents (i.e. with very high or low quantity of analyte) in the outliers [28]. In this case, it may be appropriate to include the sample in the model, as it represents a composition that could be encountered during the prediction stage. For this reason in this study we don't remove the outliers present in wine and orange juice datasets. To decrease the importance of the outliers, there are in the literature the robust regression methods which use other errors criteria [41]. It should be noted that the outliers detected in the wine and orange juice datasets will not be eliminated, in order to evaluate the robustness of our approach.

5.6.3 Data distribution between training and validation sets

Since the selection of relevant variables strongly depends on the data distribution in the training and validation sets, as seen in subsection 5.6.1, we propose a sequential method to distribute the data between training and validation sets.

Let's consider that the size of the training set is two-third (2/3) and the size of the validation set is the remaining third (1/3) of the data. This method consists in calculating the distances between all samples; thereafter these distances are arranged by ascending order. At the beginning the first three closest samples are selected, the first two samples are included in the training set and the last sample is included in the validation set. From the remaining samples, the three samples which the are closest from each other are included in both sets. The process is repeated until all samples are distributed among both sets. This distribution method ensures a homogeneous distribution of the samples in both training and validation sets.

To illustrate this method on the orange juice dataset, we first draw randomly one quarter (54 samples) of the 218 orange juice samples to form the test set. This set is used neither in training nor in validation. Thereafter, we apply this method to the remaining samples (164 samples). We split the 164 remaining orange juice samples into two sets. In this example, the first three closest samples are selected at the beginning, the first two samples are attributed to the training set, and the last sample is attributed to the validation set. The process is repeated until all 164 orange juice samples are distributed in both sets. Figure 5.29 gives a score plot for the first two principal components ($Pc1 - Pc2$) and shows respectively the distribution of the 164 orange juice samples in training and validation sets after the application of the data distribution method, and that of 54 orange juice samples used as the test set. The training, validation and test sets contain respectively 109, 55 and 54 spectra with 700 spectral variables. In this experiment, the spectra used in the training, validation and test are illustrated in

figure 5.30

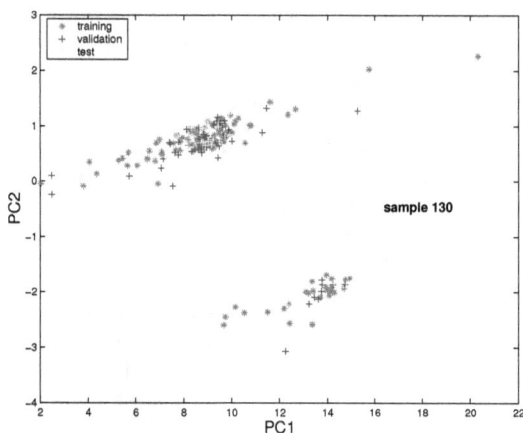

Figure 5.29: Score plot ($Pc1 - Pc2$): distribution of the orange juice samples in three sets.

In this example, we used the six methods (PCR, PLSR, FBS-lin, PCA-RBFN, PLS-RBFN and FBS-RBFN) already used in section 5.5. For the six methods in this experiment, once the best model was selected on the validation set according to a criterion (for example the number of centers in the hidden layer or the number of principal components or the number of latent variables), we concatenated the training and validation sets into a single set to estimate the parameters of the model. This model will in turn be used to estimate the normalized mean square error on the test set.

In table 5.10 the predictive ability of the six calibration models is

(a) Training set

(b) Validation set

(c) Test set

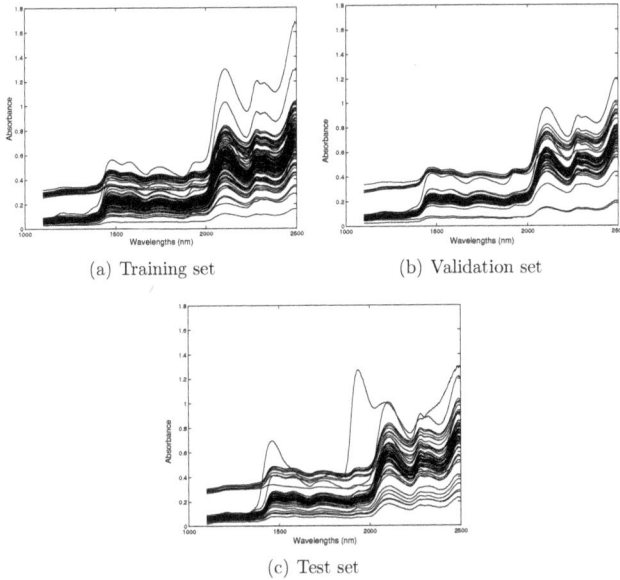

Figure 5.30: Near-infrared reflectance spectra of orange juice.

compared in terms of normalized mean square error on the training, validation and test sets.

In the PCR method, the normalized mean square error $(NMSE_T)$ on the test set was obtained with 26 principal components. Similarly for the PLSR method the $NMSE_T$ was obtained with 7 latent variables. In the FBS-lin case, the initial number of selected variables by the forward selection is 48, and this number was reduced to 38 after the application of the backward selection on the validation set, as illustrated in figure 5.31(a). In this case the $NMSE_T$ obtained on the test set is poor although the error on the validation set is smaller than the errors

Calibration model	Number of variables	$NMSE_L$	$NMSE_V$	$NMSE_T$
PCR	26	0.1263	0.2793	2.1012
PLSR	7	0.1890	0.2670	1.5299
FBS-lin	38	0.1682	0.2036	5.1122
PCA-RBFN	20	0.1340	0.2709	0.8415
PLS-RBFN	7	0.1341	0.2683	0.7958
FBS-RBFN	**14**	**0.0910**	**0.1150**	**0.6351**

Table 5.10: Results of prediction on the orange juice dataset for the six procedures on the training, validation and test sets.

obtained in PCR and PLSR methods. This is due to the bad prediction in the test stage of the sample 130, as illustrated in Figure 5.32(c). In the case of the PCA-RBFN method the $NMSE_T$ was obtained with 20 principal components and with 5 centers in the hidden layer of the RBF Network. However in the case of PLS-RBFN method the $NMSE_T$ was obtained with 7 latent variables and with 12 centers in the hidden layer of RBF Network. About the FBS-RBFN, the number of variables selected by forward selection is 14, but after the backward selection this number was not reduced, as shown in figure 5.31(b). The $NMSE_T$ was obtained with 14 relevant variables and with 15 centers (neurons) in the hidden layer of the RBF Network.

The predicted saccharose concentration according to the actual saccharose concentration on the test set with linear and non-linear calibration models are respectively presented in figures 5.32 and 5.33.

In this experiment to figures 5.32 and 5.33, and the results obtained in table 5.10, show clearly that the best model, i.e. the smallest error on the test set is obtained with our incremental procedure (FBS-RBFN), in spite of the presence of outlier in the test set (sample 130).

It should be noted that if we have a great number of samples in

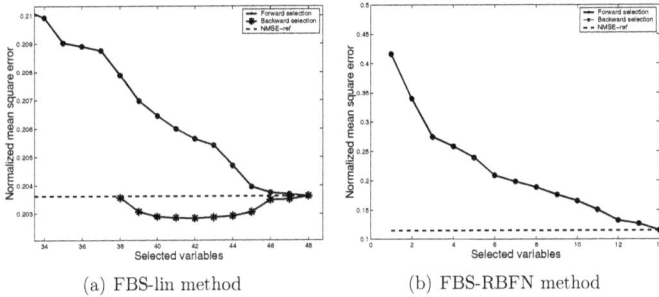

(a) FBS-lin method   (b) FBS-RBFN method

Figure 5.31: $NMSE_V$ with respect to the number of selected variables for the orange juice dataset.

the dataset, we use the random distribution of the data between training and validation sets. On the other hand, if the number of samples is small (what is often the case in practice), we use the sequential distribution method proposed in this section.

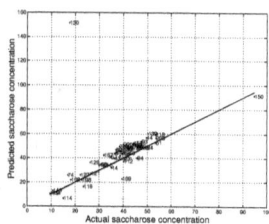

(a) PCR method

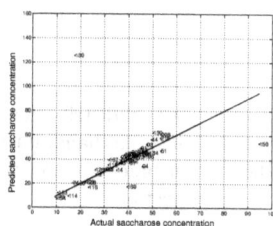

(b) PLSR method

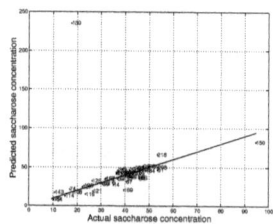

(c) FBS-lin method

Figure 5.32: Predicted saccharose concentration with respect to measured saccharose concentration on the test set with linear methods for the orange juice samples (note that the y axis scale is different in the third diagram).

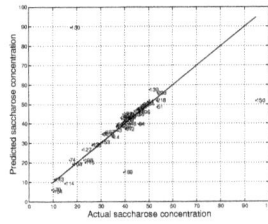

(a) PCA-RBFN method

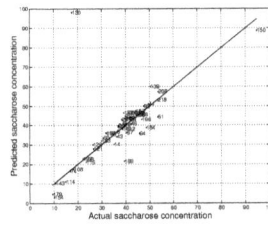

(b) PLS-RBFN method

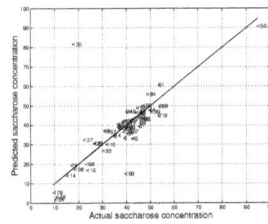

(c) FBS-RBFN method

Figure 5.33: Predicted saccharose concentration with respect to measured saccharose concentration on the test set with non-linear methods for the orange juice samples.

5.6.4   Spectrophotometric variable selection by mutual information

We proposed in section 5.4 a procedure for spectral data selection in infrared spectroscopy based on the combination of three mechanisms (nonlinear regression RBFN, incremental procedure of variable selection and use of a validation set), improving the prediction performances compared to the linear selection methods. We propose in this subsection to improve this method by a judicious choice of the first spectral data, which has a very great influence on the selection of the other variables and on the final performances of prediction. The idea is to use a measure of the mutual information between the spectral data (independent variables) and the concentration of analyte (dependent variable) to select the first variable, because this measure allows to discern the variable (spectral data) which has more meaning in relation to the analyte. Once the first variable is selected, the incremental procedure (forward-backward selection) is used to select the next spectral variables. If the idea to use a mutual information criterion for the choice of the other spectral data can also seem relevant, its implementation becomes more and more difficult when the number of selected variables increases (empty space phenomenon [74]). We will thus limit the use of the mutual information criterion to the crucial choice of the first spectral variable, and we will see in the results section that it is advantageous to combine both approaches.

5.6.4.1   Mutual information

In this subsection we will explain how mutual information can be used to assess the importance of each variable (spectral data) with respect to the calibration model.

5.6.4.2   Definition

The main goal of a prediction model is to minimize the uncertainty on the output variable. A good formalization of the uncertainty of a random variable is given by Shannon's information theory [75]. While first developed for binary variables, it has been extended to continuous variables.

The uncertainty of a random variable $\mathbf{y}$ with values v in a finite set D can be measured by means of its entropy H:

$$H(\mathbf{y}) = -\sum_{D} P(\mathbf{y} = v) \cdot \log P(\mathbf{y} = v). \qquad (5.31)$$

To illustrate this concept, let us suppose that in an extreme case all values v in D have null probability except one, which has a probability equal to 1. Then, there is absolutely no uncertainty since $\mathbf{y}$ is a constant; $H(\mathbf{y}) = 0$. Suppose now that all the values in D are equiprobables. Uncertainty is then maximal and its value is $H(\mathbf{y}) = \log N$, where N is the cardinal of D.

When the value of another variable $\mathbf{x}_i$ with values v' in D' is known, one can define conditional entropy:

$$H(\mathbf{y} \mid \mathbf{x}_i) = -\sum_{D'} P(\mathbf{x}_i = v') \sum_{D} P(\mathbf{y} = v \mid \mathbf{x}_i = v') \cdot \log P(\mathbf{y} = v \mid \mathbf{x}_i = v').$$
$$(5.32)$$

Mutual information between $\mathbf{x}_i$ and $\mathbf{y}$ is then defined by:

$$I(\mathbf{y}, \mathbf{x}_i) = H(\mathbf{y}) - H(\mathbf{y} \mid \mathbf{x}_i). \qquad (5.33)$$

The last term represents the decreasing of uncertainty on $\mathbf{y}$ when $\mathbf{x}_i$ is known.

The concepts of entropy, conditional entropy and mutual information, can be extended to the continuous case (set D of infinite size). Mutual information between variables $\mathbf{y}$ and $\mathbf{x}_i$ may be expressed by [42, 43]:

$$I = \int h(\mathbf{x}_i, \mathbf{y}) \cdot \log \frac{h(\mathbf{x}_i, \mathbf{y})}{f(\mathbf{x}_i) \cdot g(\mathbf{y})} d\mathbf{x}_i d\mathbf{y}, \qquad (5.34)$$

where $f(\mathbf{x}_i)$ and $g(\mathbf{y})$ are the marginal probability densities of variables $\mathbf{x}_i$ and $\mathbf{y}$ respectively, and $h(\mathbf{x}_i, \mathbf{y})$ is the joint probability density function of $\mathbf{x}_i$ and $\mathbf{y}$.

This formulation shows that the mutual information between $\mathbf{x}_i$ and $\mathbf{y}$ is zero if and only if $\mathbf{x}_i$ and $\mathbf{y}$ are statistically independent. The mutual information is not affected by any variable transformation, and does not make any assumption on the underlying relationship between $\mathbf{x}_i$ and $\mathbf{y}$.

5.6.4.3   Computation of the mutual information

The computation of the mutual information is based on the estimation of the marginal probability densities and the joint probabilities density function. That estimation must be carried out on the dataset. Histograms and kernels based pdf estimation are among the most commonly used [73]. In this book we used histogram estimation, because it requires less computations than kernel based estimation.

Since we need to estimate both joint and marginal probability densities, we must construct a bi-dimensional histogram. This is done using the *ith* column of $\mathbf{X}$ and the vector of measured responses (dependent variable). The procedure starts by building a bi-dimensional grid spanning the cartesian product of the domains of both variables, and then counting the number of pairs $(\mathbf{x}_i, \mathbf{y})$ that fall into a particular cell. The sizes of the cells are important parameters that have to be chosen carefully. If the cells are too large, the approximation will not be precise enough; if they are too small, most of them will be empty and the approximation will not be sufficiently smooth. Even if heuristics were proposed [17, 47, 72] to guide this choice, only the experiment can lead to an optimal choice. In our case we will limit ourselves to regular grids, making all cells the same size. Once both the joint and marginal probability densities have been estimated, the mutual information is computed using Eq. 5.34 and the fact that:

$$f(\mathbf{x}_i) = \int h(\mathbf{y}, \mathbf{x}_i)\,d\mathbf{y}, \qquad (5.35)$$

$$g(\mathbf{y}) = \int h(\mathbf{y}, \mathbf{x}_i)\,d\mathbf{x}_i. \qquad (5.36)$$

5.6.4.4   Variable selection and validation by non-linear models

The variable selection procedure which will be used in the non-linear prediction model among n available spectral variables, is then the following one [7, 8]:

1. The first selected spectral data is the one which maximizes the mutual information with the dependent variable, according to the computation detailed in the previous subsection.

2. Other spectral data are selected according to the forward-backward selection procedure by RBF Networks described in subsection 5.4.1.

It should be noted that during step 1, we only need the training set, since the computation of the mutual information does not require the estimation and the comparison of models. On the other hand for step 2, the use of other data (validation set) independent from the training set, for the computation of the $NMSE_V$ is necessary to detect and avoid the overfitting phenomenon (see figure 5.5).

5.6.4.5 Results

We applied this variable selection method to the three datasets (orange juice, milk powder and apple) already used in this chapter.

Table 5.11 shows the number of selected variables as well as the obtained errors ($NMSE_V$) by using the new method (MI+FBS-RBFN) for the three datasets. MI represents the mutual information between the dependent variable and independent variables.

Calibration model	Dataset	Number of variables	$NMSE_V$
MI+FBS-RBFN	Orange juice	36	0.0313
	Milk powder	18	0.4816
	Apple	12	0.2321

Table 5.11: Number of selected variables and their corresponding $NMSE_V$ for the three datasets.

Figure 5.34 shows the spectrum of mutual information between the independent variables and the dependent variable on the training set for the three datasets. In the case of the orange juice dataset (see figure 5.34(a)), the first variable selected by this method is the $\mathbf{x}_{690}$ variable which corresponds to the end of the spectrum (see figure 5.9(a)) where the absorbance ($\log \frac{1}{R}$) is high, i.e. where the relationship is less linear according to the hypothesis of the shortening of the optical path [57]. On the other hand the method of least squares first selects the $\mathbf{x}_{256}$ variable where the absorbance begins to become important (see figure 5.9(a)). In the case of the milk powder dataset (see figure 5.34(b)), the first selected variable by this method is $\mathbf{x}_{625}$ which corresponds to the fat in the spectra (see figure 5.15(a)). The specialists know well that there is an inverse relationship between fat and water. The variable selected by mutual information method corresponding to the fat does not suffer from the spectral interference that undergo the other variables and the other constituents in the spectra of the milk powder. In the case of the apple dataset, the first selected variable by mutual information method is $\mathbf{x}_{52}$ (see figure 5.34(c)). Concerning this last dataset we can not say anything, because we do not have any spectrochemical explanation.

About the MI-FBS-RBFN procedure, after the selection of the first variable we tested Radial Basis Function Networks with 1 to 10 centers (neurons) for the orange juice dataset, 1 to 8 centers for the milk powder dataset and 2 to 20 centers for the apple dataset in the hidden layer. The best results were obtained with respectively 5, 1 and 13 centers in the hidden layer and respectively 36, 18 and 12 selected variables for the three datasets.

Figure 5.35(a) shows the relationship between the predicted saccharose concentration and the actual concentration with the MI+FBS-RBFN variable selection method. This figure shows the improvement obtained with the use MI+FBS-RBFN procedure compared to the other calibration models, and in particular compared to the FBS-RBFN procedure (see figure 5.14(c)). Figure 5.35(b) illustrates the relationship between the predicted water and actual water content in a milk powder using

(a) orange juice dataset (b) milk powder dataset

(c) apple dataset

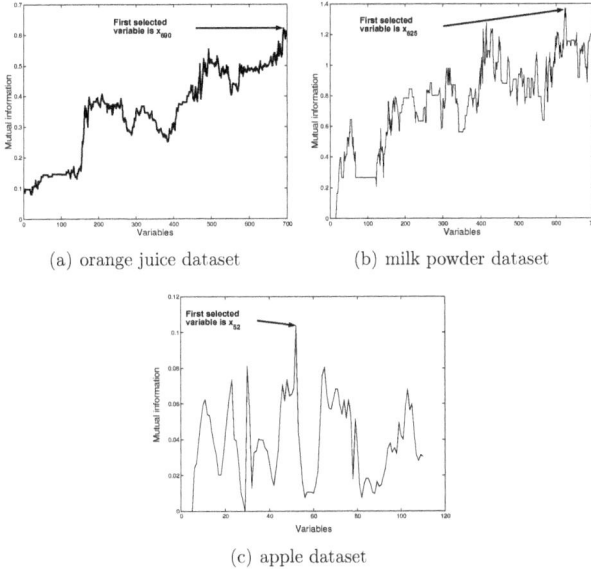

Figure 5.34: Spectrum of the mutual information between the independent variables and the dependent variable.

the MI+FBS-RBFN variable selection procedure, and the obtained improvement (see table 5.6) is shown in the same figure. Figure 5.35(c) shows the relationship between the predicted and the actual crushing force for the apple samples using the MI+FBS-RBFN method besides the improvement obtained compared to the other variable selection methods (see table 5.7).

The set of selected variables by MI+FBS-RBFN is different from the set of selected variables by FBS-RBFN procedure for each one of the three datasets. The variables obtained by MI+FBS-RBFN procedure are:

juice-subset= $\big[x_{690}\ x_{591}\ x_{552}\ x_{569}\ x_{208}\ x_{659}\ x_{590}\ x_{594}\ x_{678}\ x_{695}\ x_{519}\ x_{584}$
$x_{171}\ x_{598}\ x_{490}\ x_{491}\ x_{697}\ x_{600}\ x_{685}\ x_{566}\ x_{203}\ x_{674}\ x_{579}\ x_{658}\ x_{212}\ x_{656}\ x_{688}$
$x_{242}\ x_{677}\ x_{246}\ x_{675}\ x_{493}\ x_{495}\ x_{474}\ x_{475}\ x_{527}\big]$,

milk-subset= $\big[x_{625}\ x_{641}\ x_{636}\ x_{503}\ x_{624}\ x_{623}\ x_{640}\ x_{245}\ x_{228}\ x_{394}\ x_{493}\ x_{481}$
$x_{595}\ x_{628}\ x_{488}\ x_{600}\ x_{479}\big]$,

apple-subset= $\big[x_{52}\ x_{86}\ x_{83}\ x_{71}\ x_{68}\ x_{33}\ x_1\ x_{67}\ x_{77}\ x_{32}\ x_{37}\ x_{110}\big]$.

It should be noted that the number of selected variables by the MI+FBS-RBFN method in the three datasets (orange juice, milk powder and apple) is larger than the number of selected variables by the FBS-RBFN method (see section 5.5).

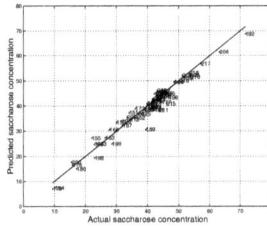

(a) orange juice dataset

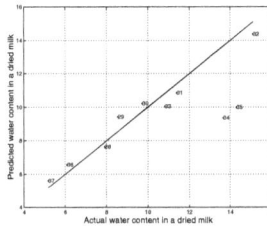

(b) milk powder dataset

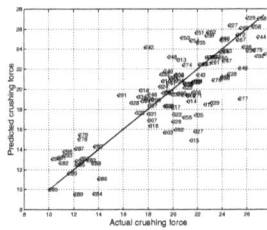

(c) apple dataset

Figure 5.35: Predicted value with respect to measured value in the three datasets with the MI+FBS-RBFN method .

Finally, we can conclude that the choice of the first selected variable has a very great influence on the performances of the final model, thus showing the interest of the combination of both approaches: the mutual information to select the first variable and the application of the FBS-RBFN method to select the next ones. According to table 5.11, we can also conclude that the MI+FBS-RBFN method gives better results than other variable selection methods (see section 5.5).

5.7 Conclusion

In this chapter, we firstly proposed a new incremental method (step by step) to select the relevant spectrophotometric variables based on the combination of three principles: linear or non-linear regression, incremental procedure for variable selection, and use of a validation set. This procedure allows on one hand to benefit from the advantages of non-linear methods to predict a chemical data, and on the other hand to avoid the overfitting phenomenon. Secondly, we proposed a sequential method to distribute data homogeneously between training and validation sets, because the variable selection methods strongly depend on the data distribution. Thirdly, we suggested a way to improve the previous method by a judicious choice of the first spectral data, which has a very large influence on the final performances of the prediction. The idea is to use the measure of the mutual information between spectral data (independent variables) and the concentration of analyte (dependent variable) to select the first variable; then the previous method (FBS-RBFN) is used to select the next spectral variables.

It should be noted that the use of mutual information to select the first variable has two advantages:

- the selected variable can have a good interpretation from the spectrochemical point of view;

- the selected variable does not depend on the data distribution in training and validation sets.

The non-linear model used in our variable selection methods (FBS-RBFN and MI+FBS-RBFN) is a RBF Network with linear and constant terms (Eq. 2.2), and is very interesting especially in the near-infrared spectroscopy, because it can handle the linear and non-linear parts of the relationship between the absorbance and the concentration of analyte. In addition, the results obtained in both proposed methods on the spectrophotometric data confirm the results obtained by Meurens [57], Achá et al. [1] and Alfaro et al. [3], concerning the linearity and the non-linearity relationship between the absorbance and concentration of analyte respectively in the mid-infrared transmittance spectra and near-infrared reflectance spectra cases.

Besides, we showed in this chapter that the best results were obtained with both our variable selection methods compared to the other calibration models used in the literature, despite the presence of the outliers in wine and orange juice datasets.

Chapter 6

Conclusion

6.1 Contributions and concluding remarks

This book explores the problem of variable selection using Artificial Neural Networks (ANN) models for high dimensional data analysis applied to mid and near-infrared spectroscopy.

Artificial Neural Networks have been invented to solve problems where other traditional data analysis tools fail. In chapter 2, we chose to focus on approximation with Radial Basis Functions Networks (RBFN). Concerning the function approximation in high dimensional spaces, we strongly believe in the advantage of RBF Networks, compared to more conventional Multi-Layer Perceptrons (MLP). RBF Networks are easier to use, their performances are less sensitive to local minima, and they do not require complex optimization algorithms.

A new method for the computation of the Gaussian function widths based on a one-dimensional exhaustive search is presented in chapter 2. With respect to fixed-width methods, this one offers the advantage of taking the distribution variations of data into account. In practice, it is able to perform much better than fixed-width methods, as it offers a greater adaptability to data.

Most criticism over Artificial Neural Networks methods over the last fifteen years is a consequence of the overfitting phenomenon encountered by many authors. It is not unusual to find articles in the literature, where authors use more parameters in their model than the number of samples available for learning; it is not difficult to obtain good learning performances in these conditions. Realistic performances, i.e. performances estimated in generalization, are however much worse. In this context, some general procedures (for example cross-validation) are widely used for model selection in order to avoid the overfitting phenomenon, and are described in chapter 3.

In chapter 4, we first show the importance of the choice of the kernel widths in RBF Networks on several artificial examples. In many situations, a bad choice can lead to an approximation error definitely higher than the optimum, sometimes by several orders of magnitudes. The comparison between the new and the state of the art methods presented in chapter 2 shows on these examples that the proposed method guarantees a natural overlap between Gaussian kernels, preserving the local properties of RBF Networks, and at the same time maximizes the generalization ability of the network.

The second part of this book is devoted to methodologies used to deal with data that can be considered as specific because of their characteristics (high-dimensional data and often non-linear relationships). The study in chapter 5 shows that datasets with a very high number of variables (several hundreds) can be treated efficiently by two new methods designed specifically for such cases.

This chapter begins by the first method (FBS-RBFN). The forward backward selection by RBF Networks is an incremental method (step-by-step) to select the relevant spectrophotometric variables based on the combination of three principles: linear or non-linear regression (RBF Networks regression), incremental procedure for variables selection, and use of a validation set. The proposed method allows on one hand to benefit from the advantages of non-linear methods (RBF Networks) to predict chemical data (that are often non-linear, especially in near-infrared spectroscopy), and on the other hand to avoid the overfitting phenomenon.

This study shows that the variable selection methods are very sensitive to the distribution of data between training and validation sets. Faced with this weakness, we propose a sequential method to distribute data homogeneously between training and validation sets.

At the end of chapter 5, we suggest to improve the incremental FBS-RBFN method by a better choice of the first spectral data, which has an important influence on the final performances of the model. The idea is to use the measure of the mutual information between spectral data and the concentration of analyte to select the first variable; then the FBS-RBFN method is used to select the next spectral data. It should be noted that the mutual information is not affected by any variable transformation, and does not make any assumption on the underlying relationship between the spectral data and the concentration of analyte. The variable selected by mutual information can have a good interpretation from the spectrochemical point of view and does not depend on the data distribution in the training and validation sets.

The results obtained with both proposed variable selection procedures on the infrared spectra (wine, orange juice, milk powder and apple datasets) confirm the results obtained by Meurens [57], Achá et al. [1] and Alfaro et al. [3], concerning the linearity and the non-linearity relationships between the absorbance and the concentration of analyte respectively in mid-infrared transmittance and near-infrared reflectance spectroscopy.

From this study, we conclude that it is not necessary to apply a non-linear model to select the relevant variables in mid-infrared transmittance spectroscopy. On the other hand, it is highly recommended to use a non-linear model for the treatment of near-infrared reflectance spectra. The non-linear model used in both our variable selection methods is a RBF Network with linear and constant terms (Eq. 2.2); it can be very interesting in the near-infrared spectroscopy, because the linear part of Eq. 2.2 can handle the linear part, and the classical

RBF Network builds the non-linear part of the model. This method provides good results for the three studied datasets (orange juice, milk powder and apple). In this study, we have also shown that the variables selected by both our approaches can have a good interpretation from the spectrochemical point of view. On the contrary, the selection methods by linear projection, such as PCR or PLSR, produce principal components or latent variables which do not have an obvious interpretation from the spectrochemical point of view.

6.2 Perspectives and future work

The variable selection using Artificial Neural Networks (RBF Networks) have been discussed and analysed thoroughly in this book. We cite here some future possible research directions.

- A measure of the mutual information between independent variables and dependent variable has been presented in order to select only the first variable. However, this measure could be used to select other variables, because it does not require the estimation and the comparison of models. The computation of the mutual information suffers nevertheless from a difficulty: it requires the estimation of the probability densities of variables. These estimations are often carried out by histograms, or by kernels. Some parameters, as the size of cells of histograms or the widths of Gaussian that are chosen, influence the quality of the probability densities estimates, and thus indirectly the relevance of the variable selection. Future research must be oriented to the comparison between the estimation of the probability densities by histograms, and the kernel methods.

- Another way to explore could be to go further into the use of mutual information on spectrophotometric data where the spectrochemical information is well established in order to better understand the choice of the selected variables.

Appendix A

In this example, we consider the adapted stack loss data given in table A.1. They consists of 17 daily observations measured in a plant for the oxidation of ammonia to nitric acid. The response variable is the percentage of ammonia lost. There are three predictor variables: x_1 is the air flow, x_2 is the temperature of the cooling water circulated through the coils in the absorption toxer for the nitric acid, and x_3 is the concentration of acid circulating. Table A.1 lists part of the stack loss data set given by Brownlee [18].

i	x_1 Rate	x_2 Temperatue	x_3 Acid concentration	y Stack loss
1	80	27	88	37
2	62	22	87	18
3	62	23	87	18
4	62	24	93	19
5	62	24	93	20
6	58	23	87	15
7	58	18	80	14
8	58	18	89	14
9	58	17	88	13
10	58	18	82	11
11	58	19	93	12
12	50	18	89	8
13	50	18	86	7
14	50	19	72	8
15	50	19	79	8
16	50	20	80	9
17	56	20	82	15

Table A.1: The adapted stack data set.

Bibliography

[1] V. Achá, H. Naveau, and M. Meurens. Extractive sampling methods to improve the sensitivity of FTIR spectroscopy in analysis of aqueous liquids. *Analusis*, 26:157–163, 1998.

[2] S. C. Ahalt and Fowler J. E. Vector Quantization using Artificial Neural Networks Models. In *Proceedings of the International Workshop on Adaptive Method and Emergent Techniques for Signal Processing and Communications*, pages 42–61, June 1993.

[3] G. Alfaro, M. Meurens, and G. S. Birth. Liquid Analysis by Dry-Extract Near-Infrared Reflectance on Fiberglass. *Applied Spectroscopy*, 44(6):979–986, 1990.

[4] A. Basilevsky. *Statistical factor analysis and related methods: theory and applications*. Willey, New York, 1994.

[5] D. Belsley, E. Kuh, and R. Welsch. *Regression diagnostics: Identifying influential data and sources of collinearity*. Wiley, New York, 1980.

[6] N. Benoudjit, C. Archambeau, A. Lendasse, J. Lee, and M. Verleysen. Width optimization of the Gaussian kernels in Radial Basis Function Networks. In *ESANN'2002*, pages 425–432, Bruges (Belgium), April 24–26, 2002.

[7] N. Benoudjit, D. François, M. Meurens, and M. Verleysen. Utilisation de l'information mutuelle pour la sélection de variables spectrales avec des modèles non-linéaires. In *Proc. Conf. Chimiométrie 2003*, pages 133–136, Paris (France), December 2003.

[8] N. Benoudjit, D. François, M. Meurens, and M. Verleysen. Spectrophotometric variable selection by mutual information. *Chemometrics and intelligent laboratory systems, Elsevier*, 74(2):243–251, December 2004.

[9] N. Benoudjit, E. Cools, M. Meurens, and M. Verleysen. Calibrage chimiométrique des spectrophotometers: sélection et validation des variables par modèles non-linéaires. In *Proc. Conf. Chimiométrie 2002*, pages 25–28, Paris (France), December 2002.

[10] N. Benoudjit, E. Cools, M. Meurens, and M. Verleysen. Chemometric calibration of infrared spectrometers: Selection and validation of variables by non-linear models. *Chemometrics and intelligent laboratory systems, Elsevier*, 70(1):47–53, January 2004.

[11] N. Benoudjit and M. Verleysen. On the kernel widths in Radial-Basis Function Networks. *Neural Processing Letters. Kluwer*, 18(2):139–154, October 2003.

[12] E. Bertran, M. Blanco, S. Maspoch, M. C. Ortiz, M. S. Sànchez, and L. A. Sarabia. Handling intrinsic non-linearity in near-infrared reflectance spectroscopy. *Chemometrics and intelligent laboratory systems*, 49:215–224, 1999.

[13] D. Bertrand and E. Dufour. *La spectroscopie infrarouge et ses applications analytiques*. Collection sciences et techniques agroalimentaires, first edition, 2000.

[14] C. M. Bishop. *Neural Networks for Pattern Recognition*. Oxford University press, 1995.

[15] M. Blanco, J. Coello, H. Iturriaga, S. Maspoch, and C. de la Pezuelo. Near infrared spectroscopy in the pharmaceutical industry. *Analyst*, 123:135R–150R, 1998.

[16] M. Blanco, J. Coello, H. Iturriaga, S. Maspoch, and J. Pagès. Calibration in non-linear near infrared reflectance spectroscopy: a comparison of several methods. *Analytica Chimica Acta*, 384:207–214, 1999.

[17] B. V. Bonnlander and A. S. Weigend. Selecting input variables using mutual information and nonparametric density estimation. In *Proceedings of International Symposium on Artificial Neural Networks (ISANN'94)*, 1994.

[18] A. Brownlee. *Statistical Theory and Methodology in Science and Engineering*. John Wiley, New York, 1965.

[19] R. A. Calvo, M. Partridge, and M. A. Jabri. A comparative study of principal component analysis techniques. In *Australian conference in Neural Networks*, Brisbane, Australia, 1998.

[20] V. Centner, O. E. Noord, and D. L. Massart. Detection of nonlinearity in multivariate calibration. *Analytica Chimica Acta*, 376:153–168, 1998.

[21] S. Chen and S. A. Billings. Neural networks for nonlinear dynamic system modelling and identification. *Int. J. Control*, 56(2):319–346, 1992.

[22] F. Despagne and D. L. Massart. Tutorial review: Neural networks in multivariate calibration. *The Analyst*, 123:157–178, 1998.

[23] N. Draper and H. Smith. *Applied regression analysis*. Wiley, New York, 1981.

[24] G. Dreyfus, J. M. Martinez, M. Samuelides, M. B. Gordon, F. Badran, S. Thiria, and L. Hérault. *Réseaux de neurones, Méthodologie et applications*. Eyrolles, France, 2002.

[25] B. Efron and R. Tibshirani. Improvements on cross-validation: The .632+ bootstrap method. *J. Amer. Statist. Assoc.*, 92:548–560, 1997.

[26] B. Efron and R. J. Tibshirani. *An introduction to the bootstrap*. Chapman and Hall, London, 1993.

[27] T. Eklove, P. Martenson, and I. Lundstrom. Selection of variables for interpreting multivariate gas sensor data. *Analytica Chimica Acta*, 381:221–232, 1999.

[28] Frédéric Estienne. *New Trends in Multivariate Analysis and Calibration*. Thesis presented to fulfil the requirements for the degree of doctor in Pharmaceutical Sciences, Vrije Universiteit Brussel, Belgium, 2002/2003.

[29] Frédéric Ferraty and Philippe Vieu. The Functional Nonparametric Model and Application to Spectrometric Data. *Computational Statistics*, 17(4), 2002.

[30] M. Forina, G. Drava, R. Boggia, S. Lanteri, and P. Conti. Validation procedures in near infrared spectrometry. *Anal. Chim. Acta*, 295:109–118, 1994.

[31] P. Geladi. Some recent trends in the calibration literature. *Chemometrics and Intelligent Laboratory Systems*, 60:211–224, 2002.

[32] P. Geladi. Review Chemometrics in spectroscopy. Part 1. Classical chemometrics. *Spectrochimica ACTA Part B*, xx:xxx–xxx, 2003.

[33] P. Geladi and E. Dabakk. An overview of chemometrics applications in NIR spectrometry. *J. NIR Spectrosc.*, 3:119–132, 1995.

[34] P. Geladi and B. R. Kowalski. Partial least squares regression: A tutorial. *Analytica Chimica Acta*, 185:1–17, 1986.

[35] Gene H. Golub and Charles F. Van Loan. *Matrix Computations*. The Johns Hopkins University Press, London, third edition, 1996.

[36] C. Goutte. Note on free lunches and cross-validation. *Neural Computation*, 9:1211–1215, 1997.

[37] A. Gresho and R. M. Gray. *Vector Quantization and Signal Compression*. Kluwer International series in engineering and computer science, Norwell, Kluwer Academic Publishers, 1992.

[38] R. Gunst and R. Mason. *Regression analysis and its applications*. M. Dekker, New York, 1980.

[39] I. Guyon and A. Elisseeff. An Introduction to Variable and Feature Selection. *Journal of Machine Learning Research*, 3:1157–1182, 2003.

[40] D. J. Hand. *Construction and Assessment of Classification Rules.* John Wiley and Sons, New York, 1997.

[41] Trevor Hastie, Robert Tibshirani, and Jerome Friedman. *The Elements of Statistical Learning, Data Mining, Inference, and Prediction.* Springer-Verlay, New York, 2001.

[42] Chi hau Chen. *Statistical Pattern Recognition.* Spartan Books., 1973.

[43] S. Haykin. *Neural Networks a Comprehensive Foundation.* Prentice-Hall, Inc., second edition, 1999.

[44] A. Hoskuldsson. PLS regression methods. *J. Chemometrics*, 2:211–228, 1988.

[45] R. J. Howlett and L. C. Jain. *Radial Basis Function Networks 2: New Avances in Design.* Physica-Verlag Heidelberg Printed in Germany, 1st edition, 2001.

[46] Young-Sup Hwang and Sung-Yang Bang. An Efficient Method to Construct a Radial Basis Function Neural Network Classifier. *Neural Networks*, 10(8):1495–1503, 1997.

[47] A. J. Izenman. Recent developments in nonparametric density estimation. *Journal of the American Statistical Association*, 86(413):205–224, 1991.

[48] Ramsay Jim and Silverman Bernard. *Functional Data Analysis.* Springer Series in Statistics. Springer Verlag, June 1997.

[49] R. Kohavi. A study of cross-validation and bootstrap for accuracy estimation and model selection. In *Proc. of the 14th Int. Joint Conf. on A.I.*, volume 2, Canada, 1995.

[50] T. Kohonen. Self-organized formation of topologically correct feature maps. *Biological Cybernetics*, 43:59–69, 1982. Reprinted in Anderson and Rosenfeld (1988).

[51] M. Kumar and D. Surya Srinivas. Unsupervised image classification by Radial Basis Function Neural Network (RBFNN). In *22nd Asian Conference on Remote Sensing*, Singapore, November 5–9, 2001.

[52] A. Lendasse, V. Wertz, and M. Verleysen. Model Selection with Cross-Validations and Bootstraps - Application to Time series Prediction with RBFN models. In *ICANN/ICONIP 13th International Conference on Artificial Neural Network and 10th International Conference on Neural Information Processing*, Istanbul, Turkey, June 26-29 2003. Artificial Neural Networks and Neural Information Processing ICANN/ICONIP 2003, O. Kaynak, E. Alpaydin, E. Oja, L. Xu eds, Springer-Verlag, Lecture Notes in Computer Science 2714, 2003, pp. 573-580.

[53] L. Ljung. *System Identification - Theory for the user*. Prentice Hall, 2nd edition, 1999.

[54] H. Martens and T. Naes. *Multivariate calibration*. Willey, Chichester, 1989.

[55] D. L. Massart, B. G. M. Vandeginste, L. M. C. Buydens, S. De Jong, P. J. Lewi, and J. Smeyers-Verbeke. *Handbook of Chemometrics and Qualimetrics : Part A*. Elsevier Science, Amsterdam, first edition, 1997.

[56] D. L. Massart, B. G. M. Vandeginste, S. N. Deming, Y. Michotte, and L. Kaufman. *Chemometrics: A textbook*. Elsevier, Amseterdam, 1988.

[57] M. Meurens. Acquisition et traitement du signal spectrophotométrique. In *La spectroscopie infrarouge et ses applications analytiques, D. Bertrand et E. Dufour*, pages 199–211. Collection sciences et techniques agroalimentaires, 2000.

[58] A. J. Miller. *Subset Selection in Regression*. CHAPMAN and HALL, London, 1990.

[59] C.E. Miller. *NIR News*, 4(6):3–5, 1993.

[60] J. Moody and C. J. Darken. Fast learning in networks of locally-tuned processing units. *Neural Computation*, 1:281–294, 1989.

[61] M. T. Musavi, W. Ahmed, K. H. Chan, K. B. Faris, and D. M. Hummels. On the Training of Radial Basis Function Classifiers. *Neural Networks*, 5:595–603, 1992.

[62] R. H. Myers. *Classical and Modern Regression with Applications*. PWS-KENT Publishing Company, second edition, 1990.

[63] S. M. Omohundro. Efficient algorithms with neural networks behavior. *Complex Systems*, 1:273–347, 1987.

[64] M. J. Orr. Introduction to Radial Basis Functions Networks, April 1996. Technical reports, www.anc.ed.ac.uk/mjo/papers/intro.ps.

[65] J. Park and I. Sandberg. Approximation and Radial Basis Function Networks. *Neural Computation*, 5:305–316, 1993.

[66] J. Park and I. W. Sandberg. Universal Approximation Using Radial-Basis-Functions Networks. *Neural Comput*, 3:246–257, 1991.

[67] T. Poggio and F. Girosi. Networks for approximation and learning. In *Proceedings of the IEEE*, volume 78, pages 1481–1497, 1990.

[68] A. C. Rencher. *Methods of multivariate analysis*. Willey, New York, 1995.

[69] B. D. Ripley. *Pattern Recognition and Neural Network*. Cambridge University Press, first edition, 1996.

[70] A. Saha and J. D. Keeler. Algorithms for Better Representation and Faster Learning in Radial Basis Function Networks. In *Advances in Neural Information Processing Systems 2*, pages 482–489. Edited by David S. Touretzky, 1989.

[71] V. D. Sanchez. Second derivative dependent placement of RBF centers. *Neurocomputing*, 7(3):311–317, 1995.

[72] D. Scott. On optimal and data-based histograms. *Biometrika*, 66:605–610, 1979.

[73] D. W. Scott. *Multivariable Density Estimation: Theory, Practice, and Visualization*. John Wiley, New-York, 1992.

[74] D. W. Scott and J. R. Thompson. Probability density estimation in higher dimension. In *Douglas, S. R. (ed): Computer Science and Statistics, Proceedings of the Fifteenth Symposium on the Interface*, pages 173–179, North Holland-Elsevier, Amsterdam, New-York, Oxford, 1983.

[75] C. E. Shannon and W. Weaver. *The Mathematical Theory of Communication*. Urbana IL: University of Illinois Press, 1949.

[76] M. A. Sharaf, D. L. Illman, and B. R. Kowalski. *Chemometrics*. Wiley, New York, 1986.

[77] G. Simon, A. Lendasse, and M. Verleysen. Bootstrap for model selection: linear approximation of the optimism. In *7th International Work Conference on Artificial Neural Networks, IWANN2003*, Mao, Menorca, (Spain), June 3-6 2003. Computational Methods in Neural Modeling, J. Mira, J.R. Alvarez eds, Springer-Verlag, Lecture Notes in Computer Science 2686, 2003, pp. I182-I189.

[78] M. Stone. An asymptotic equivalence of choice of model by cross-validation and akaike's criterion. *J. Royal. Statist. Soc.*, B39:7–44, 1977.

[79] J. M. Sutter and J. H. Kalivas. *Microchem.*, 47:60, 1993.

[80] M. Tenenhaus. *La régression PLS théorie et pratique*. Editions Technip., Paris, 1998.

[81] M. Verleysen and K. Hlaváčková. An optimized RBF network for approximation of functions. In *ESANN'94*, pages 175–180, Brussels (Belgium), April 20–24, 1994.

[82] M. Verleysen and K. Hlaváčková. Learning in RBF Networks. In *International Conference on Neural Networks (ICNN)*, pages 199–204, Washington, DC, June 3–9, 1996.

[83] Michel Verleysen. *Machine learning of high-dimensional data: Local artificial neural networks and the curse of dimensionality*. Thèse présentée en vue de l'obtention du grade d'agrégé de l'enseignement

superièur, Université catholique de louvain, Louvain-la-neuve, Belgium, December 2000.

[84] A. D. Walmsley. Improved variable selection procedure for multivariate linear regression. *Analytica Chimica Acta*, 354:225–232, 1997.

[85] Andrew Webb. *Statistical Pattern Recognition*. ARNOLD, 1999.

[86] A. S. Weigend and N. A. Gershnfeld. *Time series prediction: forecasting the future and understanding the past*. Addison-Wesley Publishing Company, Inc, 1994.

[87] S. Wold, K. Esbensen, and P. Geladi. Principal component analysis. *Chemo. Intell. lab. syst.*, 2:37–52, 1987.

[88] S. Wold, H. Martens, and H. Wold. The multivariate calibration problem in chemestry solved by the PLS method. In *Proc. Conf. Matrix pencils (A. Ruhe B. Kagstrom, eds)*, 1983. Lecture notes in Mathematics, Springer Verlag, Heidelberg, 286–293.

www.ingramcontent.com/pod-product-compliance
Lightning Source LLC
Chambersburg PA
CBHW070728220326
41598CB00024BA/3345